SpringerBriefs in Statistics

For further volumes:
http://www.springer.com/series/8921

Ali Ercan · M. Levent Kavvas
Rovshan K. Abbasov

Long-Range Dependence and Sea Level Forecasting

 Springer

Ali Ercan
M. Levent Kavvas
Department of Civil and Environmental
  Engineering
University of California
Davis, CA
USA

Rovshan K. Abbasov
Khazar University
Baku
Azerbaijan

ISSN 2191-544X   ISSN 2191-5458   (electronic)
ISBN 978-3-319-01504-0   ISBN 978-3-319-01505-7   (eBook)
DOI 10.1007/978-3-319-01505-7
Springer Cham Heidelberg New York Dordrecht London

Library of Congress Control Number: 2013946944

Printed on acid-free paper

Springer is part of Springer Science+Business Media (www.springer.com)

Contents

Chapter 1
Introduction

Abstract The stochastic approaches have been valuable in hydrological, geophysical and climatological research for representing a wide range of time series variability, uncertainty estimation, and generating future alternatives. Long-range dependence characteristics of the geophysical time series have drawn attention of scientists since Hurst phenomenon was introduced. In this study, in an effort to forecast sea levels, various statistical forecasting strategies will be discussed: ARMA (Mixed Autoregressive-Moving Average process), ARIMA (Autoregressive Integrated Moving Average process), ARFIMA (Autoregressive Fractionally Integrated Moving Average process), and trend line combined with ARFIMA (TL-ARFIMA) combination models that shall be applied to the Caspian sea level record, while applying regression to assimilate the GCM projections of sea level change to the region of Peninsular Malaysia and Malaysia's Sabah-Sarawak northern region of Borneo Island.

Keywords Long-range dependence • Long memory • ARFIMA models • Time series analysis • Sea level change • Regression techniques

Statistical and stochastic approaches are utilized extensively in applications of geophysical and climatological research to characterize and quantify spatial and temporal variability of the parameters of interest. These approaches include regression techniques (Davis 1976; Wright 1984), analysis of variance (Box et al. 1978; Cochran and Cox 1957), dimensionality reduction (Tenenbaum et al. 2000; Gamez et al. 2004), principal component analysis or Empirical Orthogonal Function analysis (Preisendorfer 1988; Von Storch and Zwiers, 1999; Jollife, 2002), Principal Oscillation Pattern analysis (Hasselman 1988; Von Storch et al. 1995; Von Storch and Zwiers, 1999), Canonical Correlation Analysis (Hotelling 1936), fractional Gaussian noise (Mandelbrot and Van Ness 1968; Mandelbrot and Wallis 1969; Mandelbrot 1971; Koutsoyiannis 2002) and autoregressive fractionally integrated moving average (ARFIMA) models (Granger and Joyeux 1980; Hosking 1981; and Geweke and Porter-Hudak 1983). The ARFIMA models are generalization of autoregressive moving average (ARMA) and autoregressive integrated moving average (ARIMA) models. Comprehensive information on ARMA and ARIMA models is provided in Box and Jenkins (1976). The stochastic approaches have been valuable in practice for representing a wide range of hyroclimatic time series variability, uncertainty estimation, and generating future alternatives (Salas et al. 1980; Beran 1994; Srikanthan and McMahon 2001; Sveinsson et al. 2003; and Koutsoyiannis 2011).

Long-range dependence or long memory characteristics of the geophysical time series have drawn attention of scientists since 1960s (Mandelbrot and Van Ness 1968; Mandelbrot and Wallis 1968, 1969) when the so called Hurst phenomenon (Hurst 1951) was discussed and explained. In addition to hydrology, long memory models have been used in several fields including astronomy, economics, and mathematics (Beran 1994). The explanation of the presence of long memory in hydrologic processes was attempted by physical mechanisms such as climate nonstationarities (Potter 1976), storage mechanisms (Klemes 1974, 1978), groundwater upwelling (Shun and Duffy 1999), and spatial aggregation (Mudelsee 2007). Long memory, that may be present in sea level records, may be due to the combination of all of the above mechanisms as the oceans are part of the earth's water cycle which is influenced by each of these mechanisms.

Sea level change has been studied by Atmosphere–Ocean coupled Global Climate Models (also called General Circulation Models) (AOGCMs) (Gregory et al. 2001; Meehl et al. 2007a) or by analyses of the historical observations of the sea level by tidal gauges (Church et al. 2004; Church and White 2006; Bindoff et al. 2007) or by satellite altimetry (Cazenave and Nerem 2004; Bindoff et al. 2007). Based on the analyses of the tidal gauge records, Church et al. (2004) determined a global mean sea level rise of 1.8 ± 0.3 mm/yr during the 1950–2000 period, and Church and White (2006) determined a mean sea level rise of 1.7 ± 0.3 mm/yr for the twentieth century. Considering these results and allowing for the upward trend in recent years by satellite altimetry observations, Bindoff et al. (2007) assessed the global mean sea level rise rate to be 1.8 ± 0.5 mm/yr for the 1961–2003 period, and 1.7 ± 0.5 mm/yr for the twentieth century.

While various authors have considered long range dependence either by means of stationary long memory models (for example, the fractional Gaussian noise model of Mandelbrot and Van Ness 1968 and Mandelbrot and Wallis 1968), or by nonstationary time trends (such as in Klemes 1974), the signal of Caspian Sea level time series seems to contain both a long term secular trend as well as long range dependent behavior. As shall be shown in the following chapters, even after removing the long term trend from Caspian Sea level time series, the residual time series still demonstrate long range dependent behavior. The example of Caspian Sea level time series demonstrates that both the long range dependence and some secular long term trend may exist together in geophysical phenomena.

In this study, in an effort to forecast sea levels, various statistical forecasting strategies will be discussed: ARMA (Mixed Autoregressive-Moving Average process), ARIMA (Autoregressive Integrated Moving Average process), ARFIMA (Autoregressive Fractionally Integrated Moving Average process), and trend line combined with ARFIMA (TL-ARFIMA) combination models that shall be applied to the Caspian sea level record, while applying regression to assimilate the GCM projections of sea level change to a particular region. The standard ARFIMA models will be applied to the annually averaged sea level observations. Finite differencing lengths for the ARFIMA models will be utilized due to the finite duration of the available observed sea level record. Sample ACFs of the residuals will be compared for various differencing lengths, and the one that minimizes the sample ACFs will be selected. Confidence intervals and the forecast updating methodology, provided for ARIMA models in Box and Jenkins (1976), will be modified for the ARFIMA models. The confidence intervals of the forecasts will be estimated utilizing the probability densities of the residuals without assuming a known distribution. ARFIMA models will also be utilized to the residuals of the linear trends; and the trend line and ARFIMA combination models will be referred to as TL-ARFIMA models. The forecasting performance of ARMA, ARIMA, ARFIMA and TL-ARFIMA models will be investigated by comparing against the observed Caspian Sea level.

Meanwhile, for the region of Peninsular Malaysia and Malaysia's Sabah-Sarawak northern region of Borneo Island, long sea level records do not exist. In such case the Global Climate Model (GCM) projections for the twenty-first century can be downscaled to the Malaysia region by means of regression techniques, utilizing the short records of satellite altimeters in this region against the GCM projections during a mutual observation period. There is substantial variability and uncertainty in the spatial distribution of sea level change among all GCMs (Meehl et al. 2007a). Climate models provide credible quantitative estimates of future climate change, particularly at continental scales and above (Randall et al. 2007). However, due to their coarse spatial grid resolution, their description of the spatial variation of the sea level change at regional and smaller spatial scales is too coarse. Therefore, a prudent projection could use the AOGCM (Coupled atmospheric-oceanic GCMs) projections for the global average sea level change, but then distribute these projections in space over regional scales according to the observed patterns based on observed sea level data by means of regression. This approach will be demonstrated for a case study along the Peninsular Malaysia and Sabah-Sarawak coastlines (Ercan et al. 2013).

The rest of this monograph is organized as follows: Long-range dependence concept is explained, methodologies developed in the literature for the estimation of long-range dependence index (Hurst Number) are provided and ARFIMA models are introduced in Chap. 2. Then, the forecasting methodology, the uncertainty estimation in the forecasts and the updating, as new data become available, are provided in Chap. 3. Afterwards, the results of the ARMA, ARIMA, ARFIMA, and TL-ARFIMA forecasting applications to the Caspian Sea level are discussed in Chap. 4. In the following chapter, the global mean sea level projections from the AOGCM simulations are assimilated to the satellite altimeter observations along Peninsular Malaysia and Sabah-Sarawak coastlines (Ercan et al. 2013). In this chapter, statistical approaches are combined with AOGCM simulation results. Conclusions drawn from each case study are provided at the end of each case study.

References

Beran J (1994) Statistics for long-memory processes. Chapman and Hall, New York

Bindoff NL, Willebrand J, Artale V, Cazenave A, Gregory J, Gulev S, Hanawa K, Le Quéré C, Levitus S, Nojiri, Y, Shum CK, Talley LD, Unnikrishnan A (2007) Observations: oceanic climate change and sea level. In: Solomon S et al. (eds) Climate change 2007: The Physical Science Basis. Contribution of working group i to the fourth assessment report of the intergovernmental panel on climate. Cambridge University Press, Cambridge, New York, pp 385-432

Box GEP, Jenkins GM (1976) Time series analysis: forecasting and control. Holden-Day, San Fransisco

Box, G.E., Hunter, W.G. and Hunter, S. (1978). Statistics for Experimenters. Wiley

Cazenave A, Nerem RS (2004) Present-day sea level change: Observations and causes. Review of Geophysics, 42(3):RG3001. doi: 10.1029/2003RG000139

Church JA, White NJ, Coleman R, Lambeck K, Mitrovica JX (2004) Estimates of the regional distribution of sea-level rise over the 1950–2000 period. J Climate 17(13):2609–2625

Church JA, White NJ (2006) A 20th century acceleration in global sea-level rise. Geophys Res Lttr 33(1):L01602. doi: 101029/2005GL024826

Cochran WG, Cox GM (1957) Experimental designs, 2nd edn. Wiley & Sons, New York

Davis RE (1976) Predictability of sea-surface temperature and sea-level pressure anomalies over the North Pacific Ocean. J Phys Oceano 6:249–266

Ercan A, Mohamad MF, Kavvas ML (2012) Sea level rise due to climate change around the Peninsular Malaysia and Sabah and Sarawak coastlines for the 21st century. Hydrol Process 27(3):367–377. doi:10.1002/hyp.9232

Gamez AJG, Zhou CS, Timmermann A, Kurths J (2004) Nonlinear dimensionality reduction in climate data. Nonlin Process Geophys 11:393–398

Geweke J, Porter-Hudak S (1983) The estimation and application of long memory time series models. J Time Ser Anal 4(4):221–238

Granger CWJ, Joyeux R (1980) An introduction to long memory time series models and fractional differencing. J Time Ser Anal 1:15–29

Gregory JM, Church JA, Boer GJ, Dixon KW, Flato GM, Jackett DR, Lowe JA, O'Farrell SP, Roeckner E, Russell GL, Stouffer RJ, Winton M (2001) Comparison of results from several AOGCMs for global and regional sea-level change 1900–2100. Clim Dyn 18(3–4):225–240

Hasselman K (1988) PIPs and POPs: the reduction of complex dynamical systems using principal interaction and oscillation patterns. J Geophys Res 93:11015–11021

Hosking JRM (1981) Fractional differencing. Biometrika 68:165–176

Hotelling H (1936) Relations between two sets of variants. Biometrika 28:321–377

Hurst HE (1951) Long-term storage capacity of reservoirs. Trans Am Soc Civil Eng 116:77–779

Jollife IT (2002) Principal Component Analysis, Springer, New York

Klemes V (1974) The Hurst phenomenon—a puzzle? Water Resour Res 10:675–688

Klemes V (1978) Physically based stochastic hydrologic analysis. Adv Hydrosci 11:285–356

Koutsoyiannis D (2002) The Hurst phenomenon and fractional Gaussian noise made easy. Hydrol Sci J 47(4):573–595

Koutsoyiannis D (2011) Hurst-Kolmogorov dynamics and uncertainty. J Am Water Resour Assoc 47(3):481–495

Mandelbrot BB (1971) A fast fractional Gaussian noise generator. Water Resour Res 7(3):543–553

Mandelbrot BB, Van Ness JW (1968) Fractional Brownian motions, fractional noises and application. Soc Ind Appl Math Rev 10:422–437

Mandelbrot BB, Wallis JR (1968) Noah, Joseph and operational hydrology. Water Resour Res 4:909–920

Mandelbrot BB, Wallis JR (1969) Computer experiments with fractional Gaussian noises. Water Resour Res 5:228–267

Meehl GA, Stocker TF, Collins WD, Friedlingstein P, Gaye AT, Gregory JM, Kitoh A, Knutti R, Murphy JM, Noda A, Raper SCB, Watterson IG, Weaver AJ, Zhao Z-C, (2007a). Global climate projections. In: Solomon S et al. (eds) Climate change 2007: The Physical Science Basis. Contribution of working group I to the fourth assessment report of the intergovernmental panel on climate change. Cambridge University Press, Cambridge, New York

Mudelsee M (2007) Long memory of rivers from spatial aggregation. Water Resour Res 43:W01202. doi:10.1029/2006WR005721

Potter KW (1976) Evidence for nonstationarity as a physical explanation of the Hurst Phenomenon. Water Resour Res 12:1047–1052. doi:10.1029/WR012i005p01047

Preisendorfer RW (1988) Principal component analysis in meteorolgy and oceanography. Elsevier, New York

Randall DA, Wood RA, Bony S, Colman R, Fichefet T, Fyfe J, Kattsov V, Pitman A, Shukla J, Srinivasan J, Stouffer RJ, Sumi A, Taylor KE (2007) Cilmate models and their evaluation. In: Solomon S et al. (eds) Climate change 2007: The Physical Science Basis. Contribution of working group I to the fourth assessment report of the intergovernmental panel on climate change. Cambridge University Press, Cambridge, New York

Salas JD, Delleur JW, Yevjevich V, Lane WL (1980) Applied modeling of hydrologic time series. Water Resources Publications, Littleton

Shun T, Duffy CJ (1999) Low-frequency oscillations in precipitation, temperature, and runoff on a west facing mountain front: a hydrogeologic interpretation. Water Resour Res 35:191–201. doi:10.1029/98WR02818

Srikanthan R, McMahon TA (2001) Stochastic generation of annual, monthly and daily climate data: a review. Hydrol Earth Syst Sci 5(4):653–670

Sveinsson OGB, Salas JD, Boes DC, Pielke RA (2003) Modeling the dynamics of long term variability of hydroclimatic processes. J. Hydrometeorol. 4(3):489–505

Tenenbaum JB, de Silva V, Langford JC (2000) A global geometric framework for nonlinear dimensionality reduction. Science 290:2319–2323

Von Storch H, Zwiers FW (1999) Statistical analysis in climate research. Cambridge University Press, Cambridge

Von Storch H, Burger G, Schnur R, von Stoch J (1995) Principal Oscillation Patterns: a review. J Climate 8:377–399

Wright PB (1984) On the relationship between indices of the southern oscillation. Mon Wea Rev 112:1913–1919

Chapter 2
Long-Range Dependence and ARFIMA Models

Abstract In this chapter, long-range dependence concept, Hurst phenomenon and ARFIMA models are introduced and the earlier work on these subjects are reviewed. Several methodologies are introduced for the estimation of long-range dependence index (Hurst number or fractional difference parameter).

Keywords Long-range dependence • Long memory • ARFIMA models • Hurst phenomenon

2.1 Long-Range Dependence

Long-range dependence or long memory has drawn the attention of scientists since 1960s when the so called Hurst phenomenon (Hurst 1951) was discussed and explained by Mandelbrot and Van Ness (1965), and Mandelbrot and Wallis (1968, 1969). Hurst (1951) investigated the water levels of Nile River for optimum dam sizing. The Hurst number, named after Harold Edwin Hurst, is an index of long memory. Hurst number H = 0 represents processes that have independent increments while 0.5 < H < 1 indicates long-range dependence.

The Hurst phenomenon has been utilized in literature extensively to asses variability of climatic and hydrologic quantities including wind power resources (Haslett and Raftery 1989), global mean temperatures (Bloomfield 1992), river flows (Eltahir 1996; Montanari et al. 1997; Vogel et al. 1998), porosity and hydraulic conductivity in subsurface hydrology (Molz and Boman 1993), indexes of North Atlantic Oscillation (Stephenson et al. 2000), tree-ring widths (Koutsoyiannis 2002), temperature anomalies in Northern Hemisphere (Koutsoyiannis 2003). In addition, long-range dependence is reported for sea levels by a power spectrum analysis (Hsui et al. 1993) and by wavelet analysis (Barbosa et al. 2006). According to Koutsoyiannis (2003), climate changes are closely related to the Hurst phenomenon, which is stochastically equivalent to a simple scaling behavior of climate variability through time.

A. Ercan et al., *Long-Range Dependence and Sea Level Forecasting*,
SpringerBriefs in Statistics, DOI: 10.1007/978-3-319-01505-7_2, © The Author(s) 2013

A process has long memory if its autocovariances are not absolutely summable (Palma 2007). The slowly decaying autocorrelations and unbounded spectral density near zero frequency are characteristics of the long memory signals (Beran 1994). A stationary process X_t has long memory (Beran and Terin 1996) if, as $|k| \to \infty$.

$$r(k) \sim L_1(k)|k|^{2H-2}, H \in (0.5,1) \tag{2.1}$$

where $r(k) = \text{cov}(X_t, X_{t+k})$ and $L_1(k)$ is a slowly varying function as $|k| \to \infty$. In other words, $L_1(ta)/L_1(t) \to 1$ as $t \to \infty$ for any $a > 0$. This implies that the correlations are not summable and the spectral density near zero frequency is unbounded. Fractional Gaussian noise (Mandelbrot and Van Ness 1968 ; Mandelbrot and Wallis 1969 ; Mandelbrot 1971, Koutsoyiannis 2002) and ARFIMA (Granger and Joyeux 1980; Hosking 1981; and Geweke and Porter-Hudak 1983) models are among the best known long memory models.

Several methodologies are available for the estimation of long-range dependence index (taken as Hurst number or fractional difference parameter which will be discussed in ARFIMA models section) such as the Rescaled Range Method (Hurst 1951; Mandelbrot and Taqqu 1979; Lo 1991), Aggregated Variance Method, Differencing the Variance Method, Absolute Moments Method, Detrended Fluctuation Analysis (Peng et al. 1994), Regression Method based on the periodogram (Geweke and Porter-Hudak 1983) and Whittle Estimator (Whittle 1951; Fox and Taqqu 1986; Dahlhaus 1989). Taqqu et al. (1995) analyzed the performance of nine different estimators. Estimation methods for long-memory models are reviewed in detail in Beran (1994), Palma (2007), and Box et al. (2008).

Minimum water levels of Nile river, as reported in Beran (1994), are depicted in Fig. 3.1 Hurst (1951) estimated the Hurst Number as 0.93 for Nile river's historical water levels. Sample ACF, periodogram, and the logarithm of the periodogram of minimum water levels of Nile River are depicted in Fig. 3.2 The calculated Hurst number and the Fig. 3.2 clearly show the long memory behavior of the historical Nile river levels.

The ARFIMA models are generalization of the linear stationary ARMA and linear nonstationary ARIMA models. The autoregressive (AR) model of first order emerged as the dominant model for background climate variability for over three decades; however, some aspects of the climate variability are best described by the long memory models (Vyushin and Kushner 2009).

2.2 ARFIMA Models

According to Beran (1994), stochastic processes may be utilized to model the behavior of observed time series solely by the statistical approach without a physical interpretation of the process parameters. Long-range memory models, ARFIMA processes in particular, have been used extensively in different fields such as astronomy, economics, geosciences, hydrology and mathematics

(Beran 1994). Early applications of the ARFIMA model were performed by Granger and Joyeux (1980), Hosking (1981) and Geweke and Porter-Hudak (1983). The process (X_t) is said to be an ARFIMA(p,d,q) process if it is a solution to the following difference equation:

$$\phi(B)(1 - B)^d X_t = \theta(B)\varepsilon_t \qquad (2.2)$$

where $\phi(z) = 1 - \sum_{j=1}^{p} \phi_j z^j$ and $\theta(z) = 1 + \sum_{k=1}^{q} \theta_k z^k$ are the autoregressive and moving-average operators, and d is the fractional difference parameter.$\varepsilon_t$is the white noise process with $E(\varepsilon_t) = 0$ and variance σ_ε^2. B is the backward shift operator such that $BX_t = X_{t-1}$. For any real number d > -1, the difference operator $(1 - B)^d$ can be defined by means of a binomial expansion (Brockwell and Davis 1987).

$$(1 - B)^d = \sum_{j=0}^{\infty} \frac{\Gamma(j - d)}{\Gamma(j + 1)\Gamma(-d)} B^j \qquad (2.3)$$

where $\Gamma(\cdot)$is the gamma function,

$$\Gamma(z) = \int_0^{\infty} t^{z-1} e^{-t} dt \text{ for } z > 0 \qquad (2.4)$$

$\Gamma(z) = \infty$ for $z = 0$, and $\Gamma(z) = z^{-1}\Gamma(1 + z)$ for $z < 0$. The process is stationary and invertible when $\phi(z)$ and $\theta(z)$ have all their roots outside the unit circle, have no common roots, and $-0.5 < d < 0.5$ (Crato and Ray 1996). The process has long memory when $0 < d < 0.5$ and is nonstationary for d $\geq$ 0.5 (Beran 1994). When d = 0, the model is referred to as autoregressive moving average (ARMA) model of order (p,q), and is capable of modeling linear stationary processes. On the other hand, when d is a positive integer, the model is referred to as autoregressive integrated moving average (ARIMA) process, and is capable of modeling linear non-stationary processes (Box and Jenkins 1976).

Box and Jenkins (1976) provide the details on the forecasting, confidence band estimation and forecast updating for ARIMA models. In this study, as described in detail in the next chapter, the governing equations, provided in Box and Jenkins (1976), are utilized and extended for ARFIMA models that use the non-integer differencing parameter d.

References

Barbosa SM, Fernandes MJ, Silva ME (2006) Long-range dependence in North Atlantic sea level. Phys A 371(2):725–731

Beran J (1994) Statistics for long-memory processes. Chapman and Hall, New York

Beran J, Terrin N (1996) Testing for a change of the long-memory parameter. Biometrika 83(3):627–638

Bloomfield P (1992) Trends in global temperature. Clim Change 21(1):1–16

Box, G.E.P, and Jenkins, G.M. (1976). Time series analysis: forecasting and control. Holden-Day, San Fransisco

Box GEP, Jenkins GM, Reinsel GC (2008) Time series analysis: forecasting and control. Wiley, Hoboken

Brockwell PJ, Davis RA (1987) Time series: theory and methods. Springer, New York

Crato N, Ray BK (1996) Model selection and forecasting for long-range dependent processes. J Forecast 15:107–125

Dahlhaus R (1989). Efficient parameter estimation for self-similar processes. Ann Stat 17, 1749–1766

Eltahir EAB (1996) El Nino and the natural variability in the flow of the Nile River. Water Resour Res 32(1):131–137

Fox R, Taqqu MS (1986) Large-sample properties of parameter estimates for strongly dependent stationary gaussian time series. The Ann Stat 14(2):517–532

Geweke J, Porter-Hudak S (1983) The estimation and application of long memory time series models. J Time Ser Anal 4(4):221–238

Granger CWJ, Joyeux R (1980) An introduction to long memory time series models and fractional differencing. J Time Ser Anal 1:15–29

Haslett J, Rafterv AE (1989) Space-time modelling with long-memory dependence: assessing Ireland's wind power resource. Appl Statist 38(1):1–50

Hosking JRM (1981) Fractional differencing. Biometrika 68:165–176

Hsui, AT., Rust, K A., Klein, G D. (1993). A fractal analysis of Quaternary, Cenozoic- Mesozoic, and Late Pennsylvanian sea level changes. J Geophys Res 98 (B12), 21963–21967

Hurst HE (1951) Long-term storage capacity of reservoirs. Trans Am Soc Civ Eng 116:77–779

Koutsoyiannis D (2002) The hurst phenomenon and fractional Gaussian noise made easy. Hydrol Sci J 47(4):573–595

Koutsoyiannis D (2003) Climate change, the Hurst phenomenon, and hydrological statistics. Hydrol Sci J 48(1):3–27

Lo A (1991) Long-term memory in stock market prices. Econometrica 59:1279–1313

Mandelbrot BB (1971) A fast fractional Gaussian noise generator. Water Resour Res 7(3):543–553

Mandelbrot BB, Van Ness JW (1968) Fractional Brownian motions, fractional noises and application. Soc Ind Appl Math Rev 10:422–437

Mandelbrot BB, Wallis JR (1968) Noah, Joseph and operational hydrology , Water Resour. Res. 4:909–920

Mandelbrot BB, Wallis JR (1969) Computer experiments with fractional Gaussian noises. Water Resour Res 5:228–267

Mandelbrot BB, Taqqu MS (1979) Robust R/S analysis of long run serial correlation. 42nd Session of the International Statistical Institute. Manila, Book 2:69–99

Molz FJ, Boman GK (1993) A fractal-based stochastic interpolation scheme in subsurface hydrology. Water Resour Res 29(11):3769–3774

Montanari A, Rosso R, Taqqu MS (1997) Fractionally differenced ARIMA models applied to hvdrologic time series. Water Resour Res 33(5):1035–1044

Palma W (2007) Long-memory time series: theory and methods. Wiley, Hoboken

Peng CK, Buldyrev SV, Havlin S, Simons M, Stanley HE, Goldberger AL (1994) Mosaic organization of DNA nucleotides. Phys Rev E 49:1685–1689

Stephenson, D B.,Pavan, V., Bojariu, R (2000). Is the North Atlantic Oscillation a random walk? Int J Clim 20(1), 1-18

Taqqu, M.S., Teverovsky, V., Willinger, W. (1995). Estimators for long-range dependence: An empirical study. Fractals 3(4), 785-798

Vogel RM, Tsai Y, Limbrunner JF (1998) The regional persistence and variability of annual streamflow in the United States. Water Resour Res 34(12):3445–3459

Vyushin DI, Kushner PJ (2009) Power-law and long-memory characteristics of the atmospheric general circulation. J Clim 22(11):2890–2904

Whittle P (1951) Hypothesis testing in time series analysis. Hafner, New York

Chapter 3
Forecasting, Confidence Band Estimation and Updating

Abstract In this chapter, forecasting, forecast confidence band estimation, and the forecast updating methodologies, provided for ARIMA models in the literature, are modified and presented for the ARFIMA models.

Keywords Forecasting, • Confidence band estimation • Updating • ARFIMA models • Confidence limits

3.1 Forecasting

Forecasting and updating of ARFIMA processes are a natural extension of those of ARIMA models. When fractional difference parameter d is 0, Eq. (2.2) represents the ARMA processes and when it is integer, Eq. (2.2) represents the ARIMA processes. Box and Jenkins (1976) provide the details on the forecasting and updating of classic ARIMA models. Forecasting as a conditional expectation of X_{t+l} is said to be made at origin t for a lead time $l \geq 1$ when written as an infinite sum of previous observations plus a random shock;

$$[X_{t+l}] = \hat{X}_t(l) = \sum_{j=1}^{\infty} \pi_j [X_{t+l-j}] + [\varepsilon_{t+l}] \tag{3.1}$$

where π weights may be obtained by equating the coefficients in

$$\phi(B)(1 - B)^d = (1 - \pi_1 B - \pi_2 B^2 - \ldots)\theta(B) \tag{3.2}$$

Because of the invertibility condition, the π weights must form a convergent series. Therefore, the forecast is dependent to an important extent only on recent past values (Box and Jenkins 1976). The variance of the forecast error $e_t(l) = X_{t+l} - \hat{X}_t(l)$ is

$$\mathrm{var}(e_t(l)) = \left[1 + \sum_{j=1}^{l-1} \psi_j^2\right] \sigma_\varepsilon^2 \tag{3.3}$$

A. Ercan et al., *Long-Range Dependence and Sea Level Forecasting*,
SpringerBriefs in Statistics, DOI: 10.1007/978-3-319-01505-7_3, © The Author(s) 2013

where σ_ε^2 is the variance of the residuals and ψ weights may be obtained by equating coefficients in

$$\phi(B)(1-B)^d \left(1 + \psi_1 B + \psi_2 B^2 + \ldots\right) = \theta(B)\ldots \qquad (3.4)$$

Equations. (3.1–3.4), which are valid for ARIMA models (Box and Jenkins 1976), are also valid for the ARFIMA models when π and ψ weights are calculated for the non-integer differencing parameter d from Eqs. (3.2) and (3.4), respectively.

π weights can be obtained by inserting the definitions of the differencing operator $(1 - B)^d$, as in Eq. (2.3), and the autoregressive and moving-average operators $\phi(B)$ and $\theta(B)$ into Eq. (3.2);

$$\left(1 - \phi_1 B - \phi_2 B^2 - \ldots - \phi_p B^p\right)\left(1 + \frac{\Gamma(1-d)}{\Gamma(-d)\Gamma(2)}B + \frac{\Gamma(2-d)}{\Gamma(-d)\Gamma(3)}B^2 + \ldots\right)$$
$$= \left(1 - \pi_1 B - \pi_2 B^2 + \ldots\right)\left(1 + \theta_1 B + \theta_2 B^2 + \ldots + \theta_q B^q\right)$$
$$\qquad (3.5)$$

ψ weights can be obtained similarly by rearranging Eq. (3.4) and using the definitions of differencing and the autoregressive and moving-average operators:

$$\left(1 + \frac{\Gamma(1+d)}{\Gamma(d)\Gamma(2)}B + \frac{\Gamma(2+d)}{\Gamma(d)\Gamma(3)}B^2 + \ldots\right)\left(1 + \theta_1 B + \theta_2 B^2 + \ldots + \theta_q B^q\right)$$
$$= \left(1 - \phi_1 B - \phi_2 B^2 - \ldots - \phi_p B^p\right)\left(1 + \psi_1 B + \psi_2 B^2 + \ldots\right)$$
$$\qquad (3.6)$$

From Eqs. (3.5–3.6), one can obtain π and ψ weights if fractional difference parameter d is known along with the autoregressive and moving-average coefficients. For example, for ARFIMA(1, d, 0) process when the autoregressive coefficient is ϕ_1, π weights can be obtained from the series

$$\pi_1 = \phi_1 - \frac{\Gamma(1-d)}{\Gamma(-d)\Gamma(2)} \text{ and } \pi_j = \phi_1 \frac{\Gamma(j-1-d)}{\Gamma(-d)\Gamma(j)} - \frac{\Gamma(j-d)}{\Gamma(-d)\Gamma(j+1)}$$
$$\text{for } j = 2, 3 \ldots \quad (3.7)$$

and ψ weights can be obtained from the series

$$\psi_0 = 1, \psi_j = \phi_1 \psi_{j-1} + \frac{\Gamma(j+d)}{\Gamma(d)\Gamma(j+1)} \text{ for } j = 1, 2 \ldots \qquad (3.8)$$

3.2 Confidence Band Estimation

From the estimation of the variance of the forecast error by Eq. (3.3), one can then estimate the confidence limits Fig. (3.1). Confidence limits can be calculated from

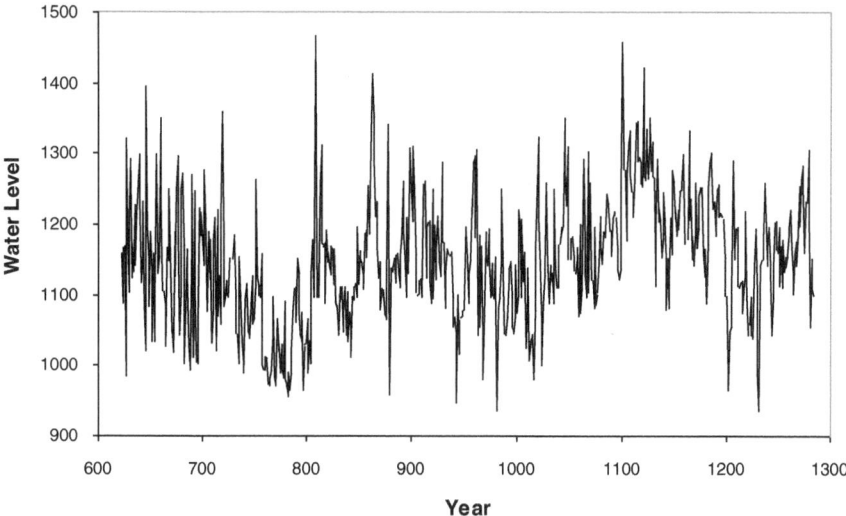

Fig. 3.1 Minimum water levels of Nile River (*source* Beran 1994)

$$X_{t+l}(\mp) = \hat{X}_t(l) + z_p^{\mp} \left[1 + \sum_{j=1}^{l-1} \psi_j^2 \right]^{1/2} \sigma_\varepsilon \tag{3.9}$$

The authors conjecture that the confidence intervals can be estimated more rigorously through calculating the standardized lower and upper bounds z_p^- and z_p^+ from the sample probability densities of the residuals. The residuals, as well as z_p^- and z_p^+, depend on the data signal and the model parameters in Eq. (2.2). Equation (3.10) below holds for the 95 % confidence limits, and one can write other confidence limits similarly.

$$P\left[z_p^- \leq \frac{\hat{X}_{t+1} - X_t(l)}{\left[1 + \sum_{j=1}^{l-1} \psi_j^2 \right]^{1/2} \sigma_\varepsilon} \leq z_p^+ \right] = 0.95 \tag{3.10}$$

The 2.5 and 97.5 percentiles of the residuals may be estimated to find z_p^- and z_p^+. If the residuals possess a normal distribution, then $z_p^{\mp} = \mp 1.96$.

One can infer from Eqs. (3.4) and (3.9) that the forecast confidence interval size depends on the variance and probability distribution of the residuals, forecast lead time l, the difference parameter d, and the autoregressive and the moving average coefficients Fig. (3.2).

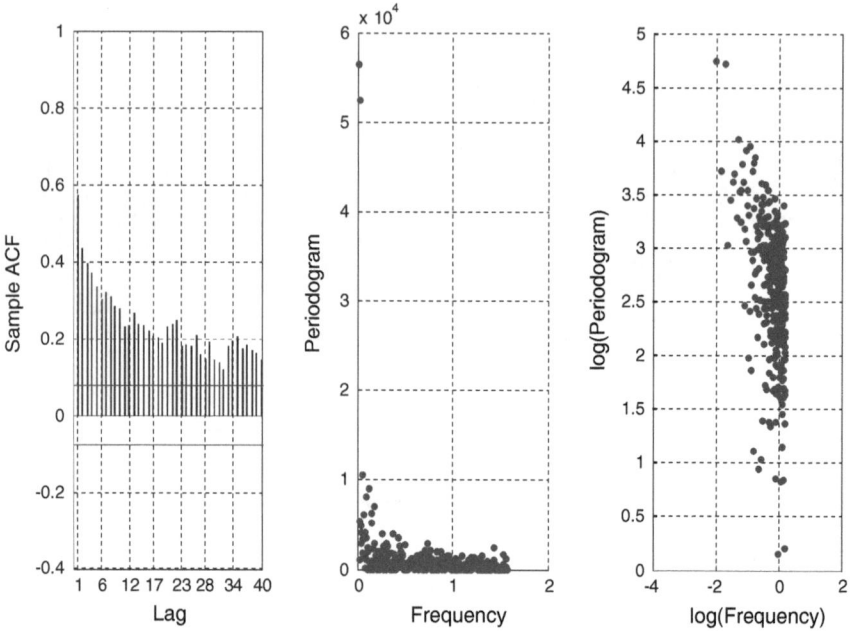

Fig. 3.2 Sample ACF, periodogram, and the logarithm of the periodogram of minimum water levels of Nile River

3.3 Updating

Updating of a forecast is important during a statistical forecasting process. However, it is usually neglected in forecast applications in the literature. Box and Jenkins (1976) provide the methodology for forecast updating for ARIMA models. Assuming that forecasts at origin t are available for lead times 1, 2, ...L, then as soon as X_{t+1} becomes available, the forecast error $\varepsilon_{t+1} = X_{t+1} - \hat{X}_t(1)$ can be calculated, and one may update the forecasts using ψ weights (Box and Jenkins, 1976) as

$$\hat{X}_{t+1}(l) = \hat{X}_t(l+1) + \psi_l \varepsilon_{t+1} \qquad (3.11)$$

for lead times 1, 2, ..., L−1. Eq. (3.11) was originally developed for the ARIMA models, but in this study it is used for the ARFIMA models as well when the ψ weights are estimated from Eq. (3.6) for the non-integer differencing parameter d.

References

Box GEP, Jenkins GM (1976) Time series analysis: forecasting and control. Holden-Day, San Fransisco

Beran J (1994) Statistics for long-memory processes. Chapman and Hall, New York

Chapter 4
Case Study I: Caspian Sea Level

Abstract The case of the Caspian Sea level time series demonstrates that both the long range dependence and some secular long term trend may exist together in geophysical phenomena. Even after removing the long term trend from the Caspian Sea level time series, the residual time series still demonstrate long range dependent behavior. Sample autocorrelation functions (ACFs) and periodograms of the sea level data are investigated and the Hurst coefficients are estimated for various time intervals. Forecasting performance of linear stationary (autoregressive moving average, ARMA), linear nonstationary (autoregressive integrated moving average, ARIMA), long-range memory (autoregressive fractionally integrated moving average, ARFIMA) and Trend Line-ARFIMA (TL-ARFIMA) combination models are investigated by comparing the forecasts with the observed Caspian Sea levels. Forecasts and their confidence bands, estimated by the ARFIMA and TL-ARFIMA models, are compared with the forecasts of the AOGCMs reported in the literature. In this study, the forecast confidence bands and the forecast updating methodology, provided for ARIMA models in the literature, are modified for the ARFIMA models. Sample ACFs are utilized to estimate the differencing lengths of the ARFIMA models. The confidence bands of the forecasts are estimated using the probability density functions of the residuals without assuming a known distribution.

Keywords ARFIMA models • Trend Line-ARFIMA (TL-ARFIMA) models • Caspian Sea Level

4.1 Introduction

The Caspian Sea is the biggest inland body of water in the world. This massive lake is located inside the Eurasian continent where the South-Eastern Europe borders with Asia. The Caspian Sea has no connection to the world's oceans, and its surface level elevation at the moment is around -27 m (27 m below the average

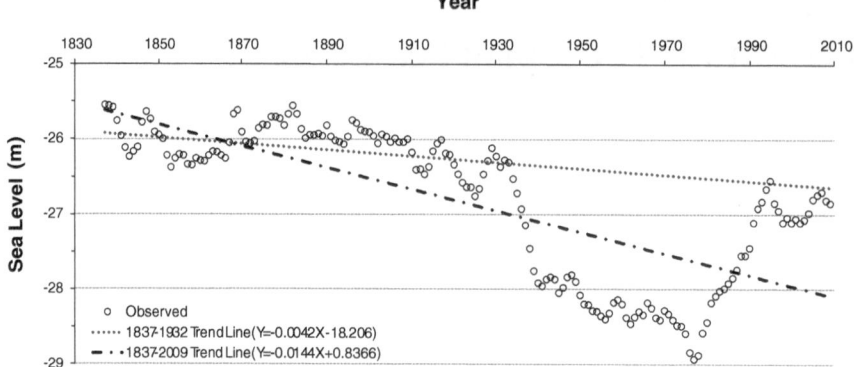

Fig. 4.1 Caspian Sea level (m) at Baku during years 1837–2009

ocean level). Its surface area is around 390,000 km^2, which is about 18 % of the surface area of all lakes in the world (Rodionov 1994).

The Caspian Sea has been under increased attention due to its unique natural characteristics and the important role it plays in the social, economic and ecological characteristics of the surrounding countries: Azerbaijan, Iran, Kazakhstan, Russia and Turkmenistan (Rodionov 1994). Exploration of large oil and gas resources increased the importance of the region over the last decades.

The Caspian Sea level observations since 1837 are depicted in Fig. 4.1. Between 1837 and 1931, the sea level fluctuated between −25.5 and −26.8 m. The sea level dropped dramatically by 1.7 m in the next nine years. In 1977, Caspian Sea reached its lowest recorded level, −29 m, and started to rise. The rise after 1977 was thought to be temporary and Kara-Bogaz-Gol Strait was dammed in 1980. In 1984, limited amount of water was allowed to flow into Kara-Bogaz-Gol Bay for restoration purposes in the Bay, and the dam was eliminated in 1992. In 1995, the Caspian Sea level reached its maximum elevation, −26.5 m, since the early twentieth century with a total rise of 2.4 m since 1977. The sea level has been stable and fluctuating slightly around −27 m in the last decade.

Caspian Sea filters climatic noise by its large surface area and volume of water. Therefore, the observed variation in the Caspian Sea level may be a good indicator of the change in climate (Rodionov 1994). Caspian Sea level fluctuations are due to a combination of many factors including climate-induced changes in the hydrological budget (Rodionov 1994; Elguindi and Giorgi 2006), neotectonic movements (Vdovykin 1990), surface-groundwater interaction (Shilo 1989), mud volcanism (Bobrow 1961), and anthropogenic activities such as land use change and change in water use in rivers draining into Caspian Sea (Rodionov 1994). Climate induced factors dominate the sea level fluctuations as anthropogenic activities are of secondary importance (Rodionov 1994).

Elguindi and Giorgi (2006) assessed possible Caspian Sea level changes in the twenty-first century using the output from seven Atmosphere-Ocean coupled General Circulation Models (AOGCMs) under A1b and A2 emission scenarios.

They found that the simulated sea level ranges between -20 and -45 m below sea level at the end of twenty-first century in both emission scenarios. Two of the models in A2 scenario and one in A1b scenario predict an increase in sea level. However, the ensemble mean of the models by the end of the twenty-first century is around -34 m below sea level, which is more than 5 m below the minimum recorded level since 1837 and is very close to the minimum observed level in the last 25 centuries. According to paleo-reconstructions of the Caspian Sea level in the last 25 centuries, the Caspian Sea level varied between -22 and -35 m below sea level (Golitysn 1995). Renssen et al. (2007) estimated a 4.2 m drop in Caspian Sea level in the twenty-first century as a result of the anthropogenic A1b scenario. It is claimed that loss in the twenty-first century overwhelms the increase in river runoff, causing the sea level drop (Elguindi and Giorgi 2006; Renssen et al. 2007).

In the following sections, ARMA, ARIMA, ARFIMA and TL-ARIMA forecasting applications to the Caspian Sea level will be discussed. Updating of a forecast will be shown to be an important part of the forecasting process. Adaptation of the ARFIMA models to sudden observed sea level variations will be compared to those of ARMA and ARIMA models applied to the Caspian Sea level data.

4.2 ARMA and ARIMA Model Forecasts

First, purely statistical forecasts were performed using the observed Caspian Sea level data from year 1837 to years 1932, 1977, 1995 by means of linear stationary and non-stationary models, and their forecast performance was evaluated by means of the observations in the following years. The AR model of first order has been utilized extensively to model background climate variability for over three decades (Vyushin and Kushner 2009). First order AR model is utilized initially for the forecasting, but the AR coefficients are estimated as unity (~0.9999) by the least square estimation during all application periods. This result means that the chosen AR model is actually equivalent to ARIMA(0,1,0) model. Hence, next the ARIMA(1,1,0) model was utilized for periods 1837–1932, 1837–1977, and 1837–1995, and the coefficients of the first order AR component of the ARIMA model were estimated as 0.3600, 0.4481 and 0.4862, respectively. The sample probability density functions of the residuals for ARMA(1,0) and ARIMA(1,1,0) models for the three time intervals are shown in Fig. 4.2. The means of the residuals are -0.010 m, -0.027 m, and -0.009 m for ARMA(1,0) models and -0.005 m, -0.014 m, and -0.003 m for ARIMA(1,1,0) models for the time intervals 1837–1932, 1837–1977, and 1837–1995, respectively. Using Eq. (3.10) and model residuals shown in Fig. 4.2, z_p^- were estimated as -1.74, -1.83, and -1.81 for ARMA models, and -1.70, -1.73, and -1.76 for ARIMA models for 1837–1932, 1837–1977, and 1837–1995 periods, respectively. Similarly, z_p^+ was estimated as 1.97, 1.90, and 2.22 for ARMA models, and 2.09, 1.84, and 2.27 for ARIMA models, respectively.

(a)   **(b)**

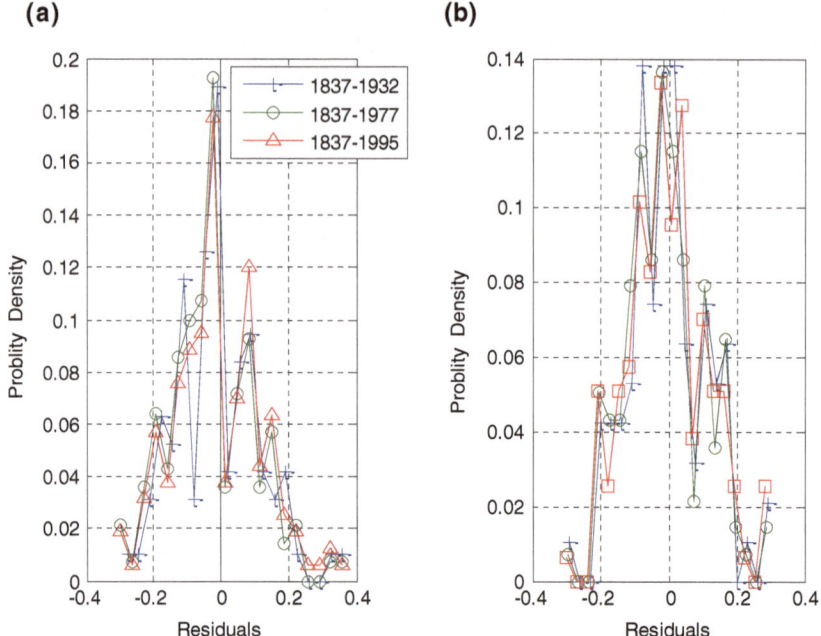

Fig. 4.2 Sample probability density functions of the residuals for **a** ARMA(1,0) models for years 1837–1932, 1837–1977, and 1837–1995 and **b** ARIMA(1,1,0) models for years 1837–1932, 1837–1977, and 1837–1995

In order to check the reliability of the developed models, Box and Jenkins (1976) suggests portmanteau lack of fit and cumulative periodogram tests as model diagnostic tools. If the fitted model is appropriate, the portmanteau lack of fit test suggests that the statistic

$$Q = n \sum_{k=1}^{K} r_k^2 \qquad (4.1)$$

for the first K autocorrelations r_k from any ARIMA(p,d,q) process is approximately Chi square distributed with (K − p − q) degrees of freedom as χ^2(K − p − q) for n = N − d, where N is the length of the residuals. Ljung and Box (1978) proposed a modified version of this statistic (modified Ljung-Box-Pierce statistic) as

$$Q = n(n+2) \sum_{k=1}^{K} (n-k)^{-1} r_k^2 \qquad (4.2)$$

On the other hand, the autocorrelations may not adequately take into account the periodic characteristics of the series (Box and Jenkins 1976). Bartlett (1955) showed that the cumulative periodogram provides an effective means for the

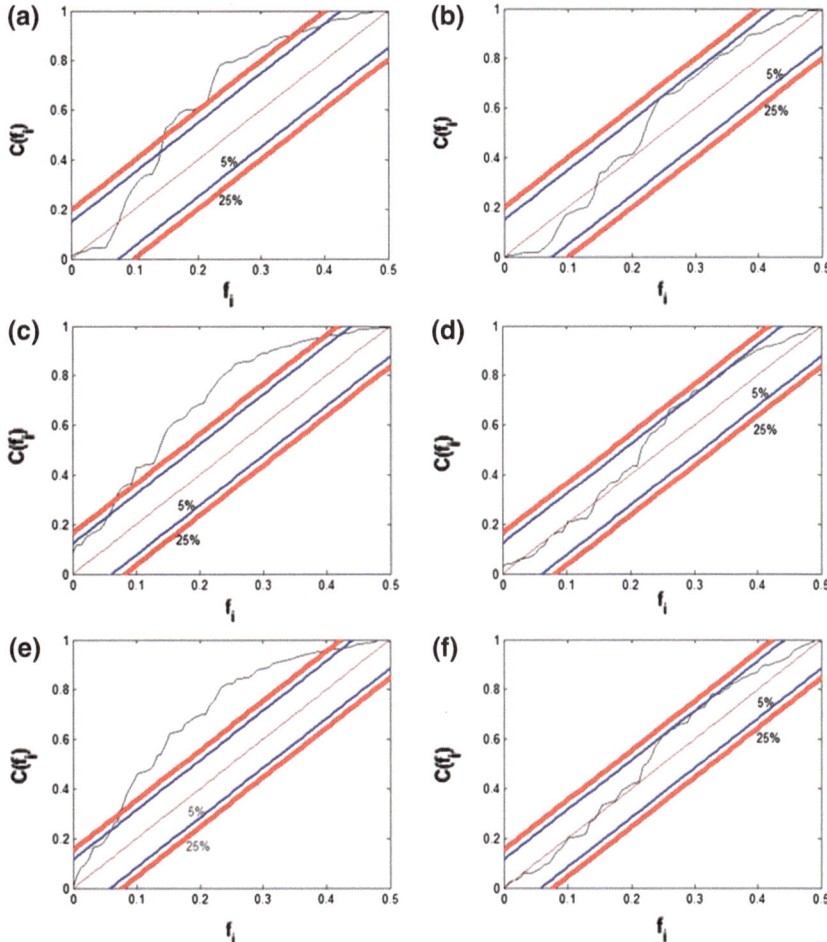

Fig. 4.3 Cumulative periodograms of residuals for **a–c** ARMA(1,0) model for years 1837–1932, 1837–1977, and 1837–1995, respectively and **d–f** ARIMA(1,1,0) model for years 1837–1932, 1837–1977, and 1837–1995, respectively

detection of periodic non-randomness. For a white noise series, the plot of the cumulative periodogram against frequency would form a straight line, joining points (0,0) and (0.5,1).

Diagnostic checks as explained in Box and Jenkins (1976) were then applied to the residuals of the developed ARMA(1,0) and ARIMA(1,1,0) models of the Caspian Sea level. For the three time intervals tested, the first order AR Models (ARMA(1,0)) did not pass at the 5 % χ^2 significance level in portmanteau lack of fit tests as applied to the first 20 sample autocorrelations of the residuals. Also, ARMA(1,0) did not pass the 5 % or 25 % Kolmogorov-Smirnov significance

Table 4.1 Summary of ARMA, ARIMA and ARFIMA models that were applied to Caspian Sea level Time Series

Model type	Model equation	Duration	Updated years	Validation period	RMSE
ARMA	$(1 - 0.9999B)X_t = \varepsilon_t$	1837–1932	–	1933–1964	1.69
		1837–1932	1933–1934	1935–1964	1.50
ARMA	$(1 - 0.9999B)X_t = \varepsilon_t$	1837–1977	–	1978–2009	1.63
		1837–1977	1978–1979	1980–2009	1.36
ARMA	$(1 - 0.9999B)X_t = \varepsilon_t$	1837–1995	–	1996–2009	0.43
		1837–1995	1996–1998	1999–2009	0.21
ARIMA	$(1 - 0.36B)(1 - B)X_t = \varepsilon_t$	1837–1932	–	1933–1964	1.69
		1837–1932	1933–1934	1935–1964	1.34
ARIMA	$(1 - 0.4481B)(1 - B)X_t = \varepsilon_t$	1837–1977	–	1978–2009	1.76
		1837–1977	1978–1979	1980–2009	1.19
ARIMA	$(1 - 0.4862B)(1 - B)X_t = \varepsilon_t$	1837–1995	–	1996–2009	0.51
		1837–1995	1996–1998	1999–2009	0.35
ARFIMA	$(1 - 0.9999B)(1 - B)^{0.45}X_t = \varepsilon_t$	1837–1977	–	1978–2009	3.09
		1837–1977	1978–1979	1980–2009	1.37
ARFIMA	$(1 - 0.9525B)(1 - B)^{0.49}X_t = \varepsilon_t$	1837–1995	–	1996–2009	1.74
		1837–1995	1996–1998	1999–2009	0.32
ARFIMA	$(1 - 0.9773B)(1 - B)^{0.499}X_t = \varepsilon_t$	1837–2009	–	–	–

levels (Hald 1952) in the cumulative periodogram tests. On the other hand, ARIMA(1,1,0) models for the three application time periods passed both the portmanteau lack of fit and the cumulative periodogram tests. For the ARIMA models of the three time periods, the Q statistics (28.5, 19.4 and 16.7) and the modified Ljung-Box-Pierce statistics (32.4, 20.8 and 17.8) are smaller than the 5 % significance level for χ^2 statistic with 19 degrees of freedom (which is 32.9). Q statistics are also smaller than χ^2 with 19 degrees of freedom (i.e. 30.1) at the 5 % significance level. The estimated cumulative periodograms against the frequency are illustrated in Fig. 4.3a–f for the residuals of ARMA and ARIMA models for the time periods 1837–1932, 1837–1977, and 1837–1995. The estimated cumulative periodograms are outside the 5 % or 25 % Kolmogorov-Smirnov limits for ARMA(1,0) models as shown in Fig. 4.3a–c. However, they are within 5 % and 25 % Kolmogorov-Smirnov limits for ARIMA(1,0) models as depicted in Fig. 4.3d–f.

Although, ARMA models did not pass the model diagnostic checks, as discussed above, we wish to discuss the difference in the forecasts of ARMA and ARIMA models. The list of ARMA and ARIMA models that were applied to Caspian Sea level and their respective Root Mean Square Error (RMSE) values, calculated in the forecast validation periods, are tabulated in Table 4.1. RMSE values are calculated with respect to the observed sea levels. ARMA(1,0) and ARIMA(1,1,0) model forecasts and 95 % confidence intervals for the Caspian Sea level after 1932, 1977 and 1995 are illustrated in Fig. 4.4a, b. Forecasts of both models after 1932, 1977 and 1995 are quite similar because the RMSE values

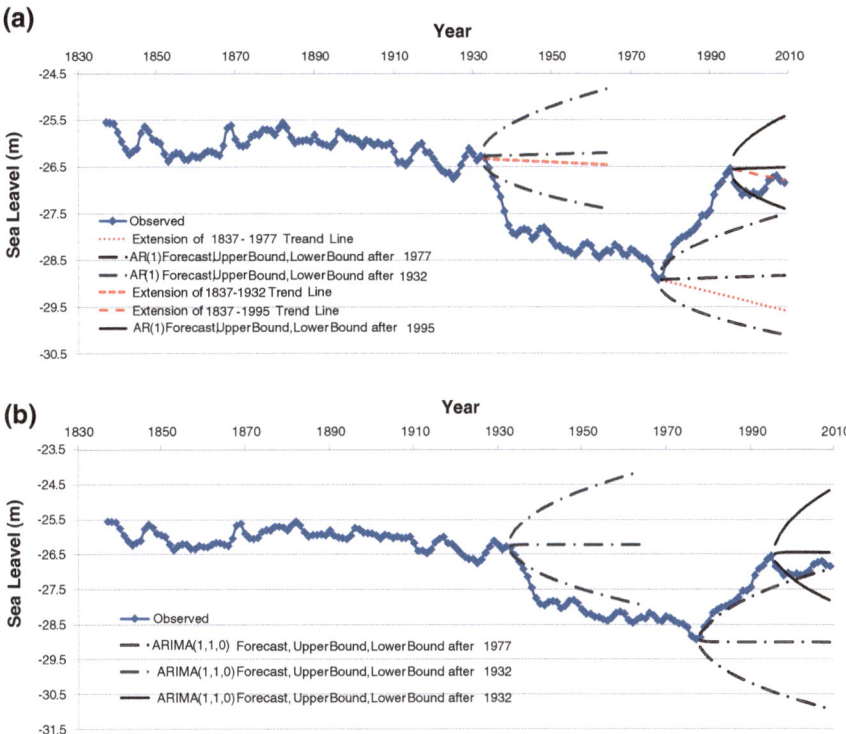

Fig. 4.4 **a** ARMA(1,0) (the *top* figure); **b** ARIMA(1,1,0) (the *bottom* figure) forecasts and 95 % confidence intervals of Caspian Sea level (m) after 1932, 1977 and 1995

are 1.69 m for both models after 1932; 1.63 m for ARMA model and 1.76 m for ARIMA model after 1977; and 0.43 m for ARMA model and 0.51 m for ARIMA model after 1995. The linear trend analyses were then performed for 1837–1932, 1837–1977, and 1837–1995 time intervals. The corresponding forecasts by extending the slopes of the linear trends after 1932, 1977, and 1995 are plotted in Fig. 4.4a. The high RMSE values for the forecasts after 1932 and 1977 confirm that both models are not quite capable of forecasting the sudden sea level changes after these years. After 1932 and 1977, neither the ARMA model nor the ARIMA model performed well because the Caspian Sea level deviated significantly from the linear trends. After 1995 both model performances improve because the sea level fluctuates around a linear trend. As depicted in Fig. 4.4a, b, the confidence intervals of the ARIMA models for forecasts after 1932, 1977 and 1995 are larger when compared to those of ARMA models. However, the observed values after 1932 and 1977 are outside the confidence intervals for both models.

The resultant forecasts and the confidence bands after 1932, 1977, and 1995 of both models when updated by Eq. (3.11), and the data of 1933–1934, 1978–1979, and 1996–1998, respectively, are shown in Fig. 4.5. The updating process results in better forecasting during the validation period for the ARMA and ARIMA models. For ARMA models, the RMSE values for the forecasts after 1934, 1979 and

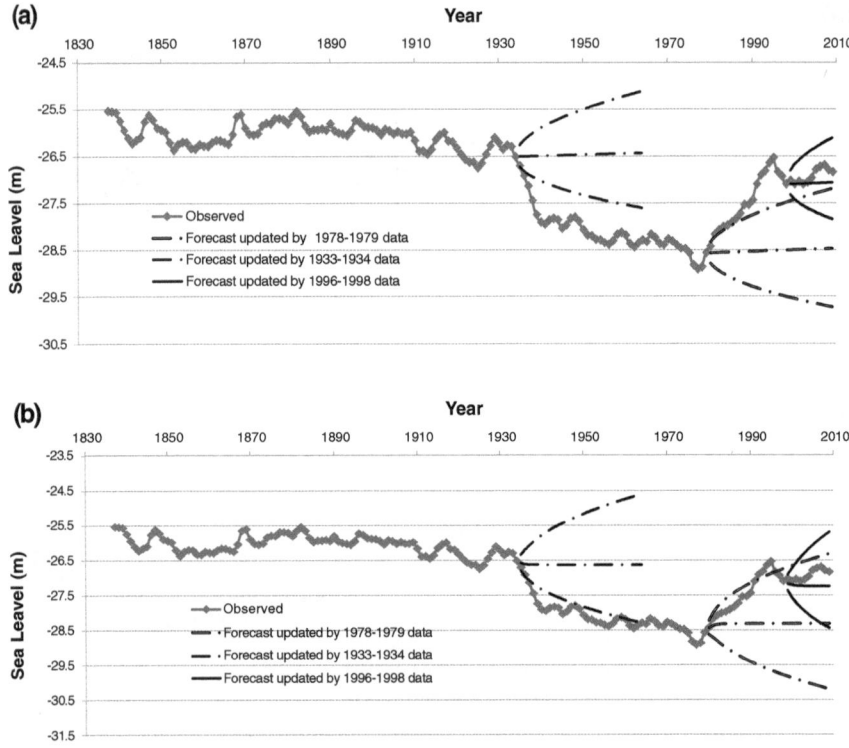

Fig. 4.5 Caspian Sea level (m) forecasts after 1932, 1977 and 1995 that are updated by data of 1933–1934, 1978–1979, and 1996–1998, respectively, using **a** ARMA(1,0) (the *top* figure), and **b** ARIMA(1,1,0) (the *bottom* figure) models

1998 (i.e. 1.50 m, 1.36 m, and 0.21 m, respectively) are lower than those before updating (i.e. 1.69 m, 1.63 m, and 0.43 m, respectively). Similarly for ARIMA models, the RMSE values for the forecasts after 1934, 1979 and 1998 (i.e. 1.34 m, 1.19 m, and 0.35 m, respectively) are lower than those before updating (i.e. 1.69 m, 1.76 m, and 0.51 m, respectively).

Although the updating resulted in smaller RMSE values, i.e. better forecasting, the updating of ARMA and ARIMA models did not capture the trend of the observed sea levels. Between 1933 and 1941, the observed Caspian Sea level dropped with an average rate of 0.21 m/year. Between 1977 and 1995 it rose with an average rate of 0.14 m/year. The forecasted sea level rise rate between 1933 and 1941 is 0.0026 m/year for the ARMA model and 0.0023 m for the ARIMA model. The updated sea level change rate between 1935 and 1941 is still 0.0026 m/year for the ARMA model and, with a slight improvement, −0.0074 m/year for the ARIMA model. Here, the negative rate represents a sea level drop. On the other hand, the forecasted sea level change rate between 1978 and 1995 is 0.0029 m/year using the ARMA model and −0.0019 m using the ARIMA model. The

updated sea level change rate between 1980 and 1995 is still 0.0029 m/year for
the ARMA model and, with a slight improvement, 0.0073 m/year for the ARIMA
model. Hence, updating of the ARMA model forecasts did not improve the performance, and updating of the ARIMA model forecasts slightly improved the forecast performance for the Caspian Sea level fluctuations and their observed trends.
Comparing the observed sea level trends against the model forecasts through
1933–1941 and 1978–1995, both the ARMA and the ARIMA models are far from
adjusting to these trends during these periods.

4.3 ARFIMA Model Forecasts

A series having a slowly declining Autocorrelation Function (ACF) or infinite spec-
tral value at zero frequency are features of long memory (Beran 1994). Sample
ACFs, periodograms and logarithms of the periodograms of the Caspian Sea level
during the periods 1837–1932, 1837–1941, 1877–1977, 1877–1995, and 1837–2009
are shown in Figs. 4.6a–e, 4.7a–e, and 4.8a–e, respectively. Sample ACFs, periodograms and logarithms of the periodograms, after removing the linear trends (i.e.
for the residuals after removing the simple linear regression line for each time interval) for the same above periods, are plotted in Figs. 4.6f–j, 4.7f–j, and 4.8f–j, respectively. The historical sea level data signal clearly possesses long range dependence
since 1977 (Figs. 4.6c–e, 4.7c–e, and 4.8c–e). Also, the residuals after removing the
linear trends, still inherit the long memory property (Figs. 4.6h–j, 4.7h–j, 4.8h–j).

  Forecasting and updating by ARFIMA models are similar to those of ARIMA
models except that the fractional difference parameter d is a real number instead of
an integer. In this study, the Hurst numbers were estimated by the Rescaled Range,
Aggregated Variance and Absolute Moments Methods. Then the average of the
three estimates was used to calculate the fractional difference parameter d in the
ARFIMA models (d = H − 0.5, where H is the Hurst number). For 1837–1977,
1837–1995, and 1837–2009 periods, d is estimated as 0.450, 0.490, 0.499, respectively, showing stationary long range behavior since $0 < d < 0.5$. However, the values of d through time indicate that Caspian Sea level is in a transition stage from
stationary long range behavior to a non-stationary behavior as d = 0.5 is the limit
of stationarity.

  The forecasting application of long-range models, ARFIMA model in particular, to physical time series is challenging because forecasting requires an expansion of the infinite series as in Eq. (3.1). For a finite series, Eq. (3.1) takes the form

$$[X_{t+l}] = \hat{X}_t(l) = \sum_{j=1}^{L} \pi_j \left[X_{t+l-j} \right] + [\varepsilon_{t+l}] \tag{4.3}$$

where L is the differencing length. In order to estimate the differencing length L
for the Caspian Sea level data, the sample ACFs of the residuals were compared
for various differencing lengths L for periods from 1837 to 1977, to 1995 and to

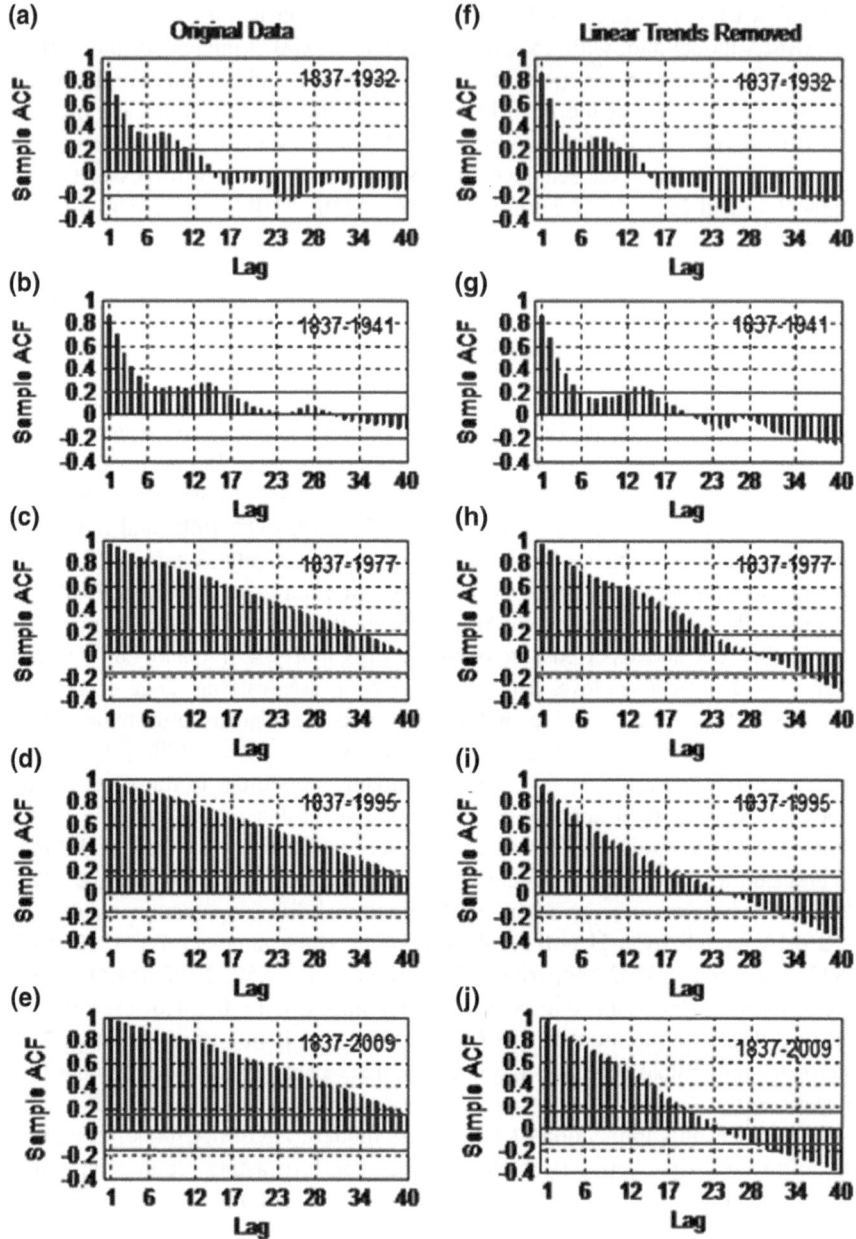

Fig. 4.6 Sample ACFs of Caspian Sea level (m) during years 1837–1932, 1837–1941, 1877–1977, 1877–1995, and 1837–2009 (**a–e**) and those after the linear trends are removed (**f–j**)

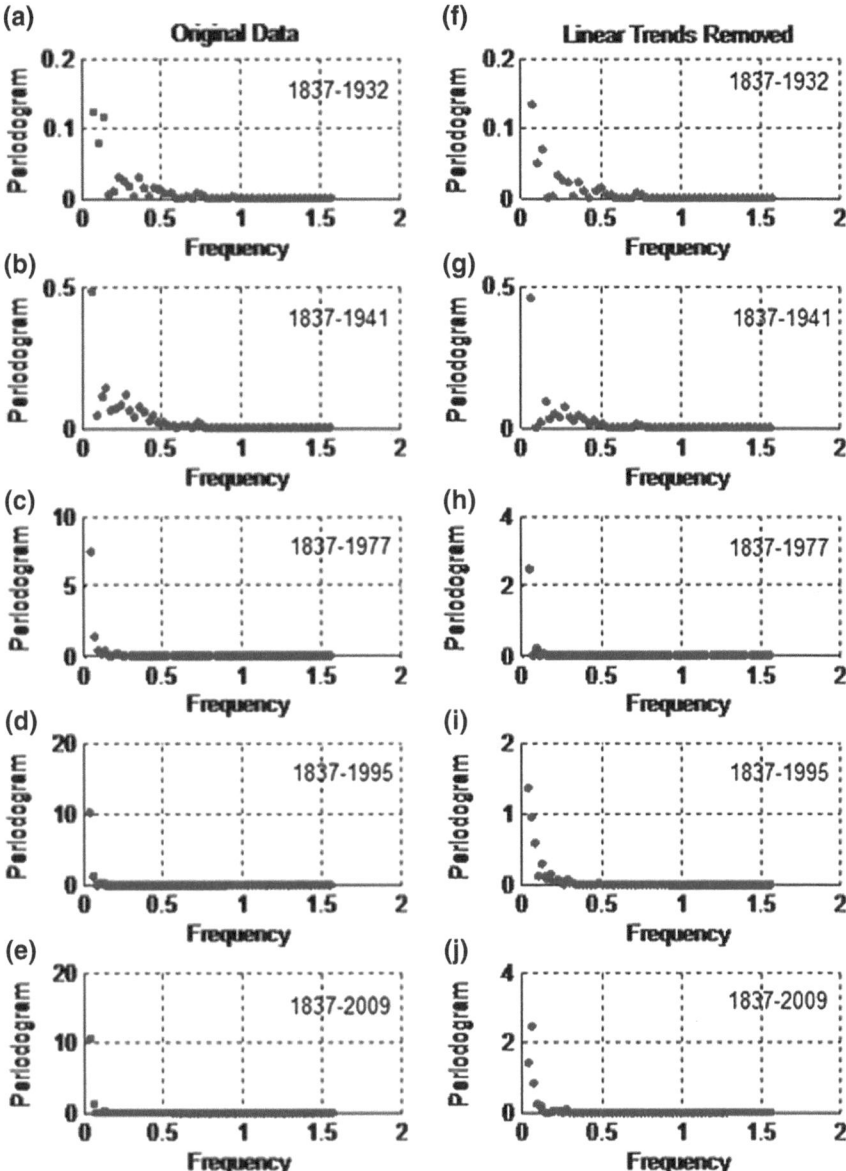

Fig. 4.7 Periodograms of Caspian Sea level (m) during years 1837–1932, 1837–1941, 1877–1977, 1877–1995, and 1837–2009 (**a–e**) and those after the linear trends are removed (**f–j**)

2009, respectively. For illustration the sample ACFs of the residuals, together with two standard error limits $(2/\sqrt{N})$, are depicted for various differencing lengths (L = 40, 60, 80 100, 120, 130, 135, 140, 145 years) for 1837–2009 period in Fig. 4.9. The differencing lengths that minimize the magnitude of the sample ACF of the residuals are estimated as 110 years for 1837–1977 period, and 140 years for

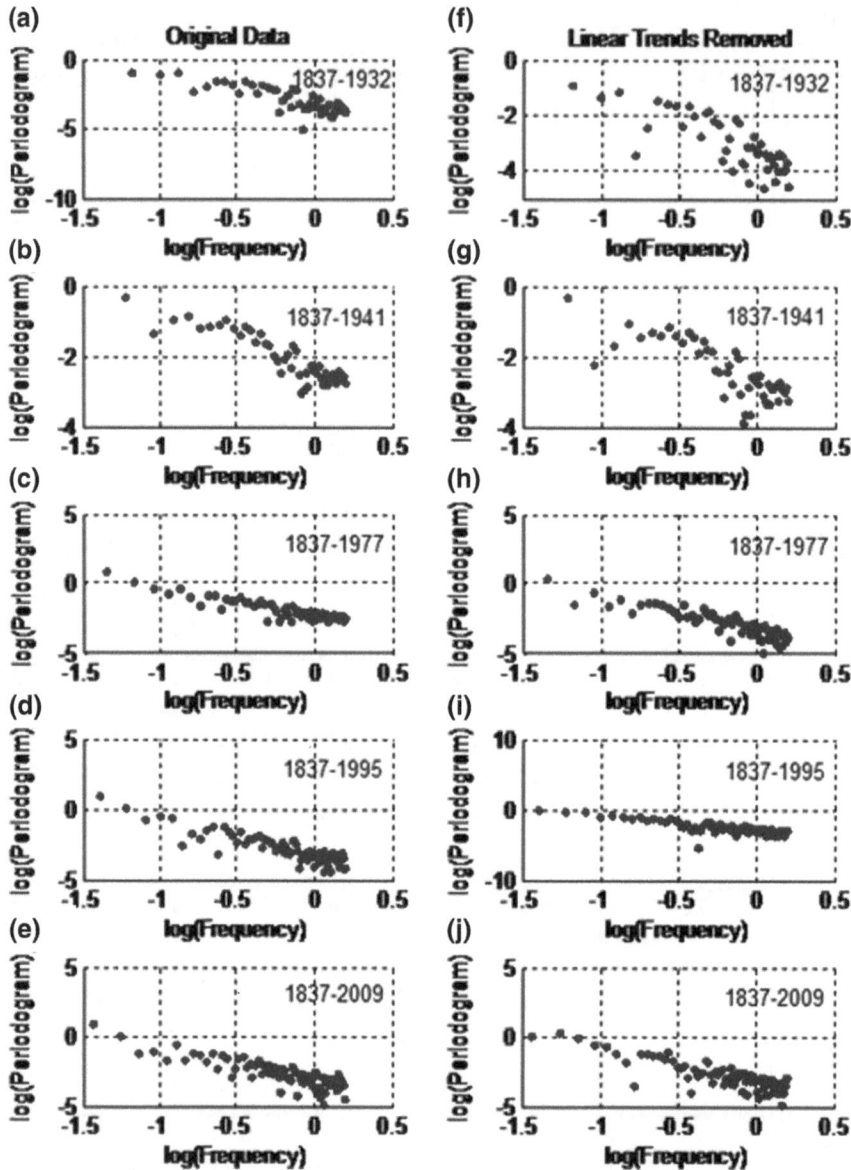

Fig. 4.8 Logarithms of periodograms of Caspian Sea level (m) during years 1837–1932, 1837–1941, 1877–1977, 1877–1995, and 1837–2009 (**a–e**) and those after the linear trends are removed (**f–j**)

1837–1995 and 1837–2009 periods. After applying fractional differencing, a first order AR model was fitted by the least squares estimation method to the residuals. The AR coefficients are estimated as 0.9999, 0.9525 and 0.9773 for 1837–1977, 1837–1995, and 1837–2009 periods, respectively. The standardized lower bound

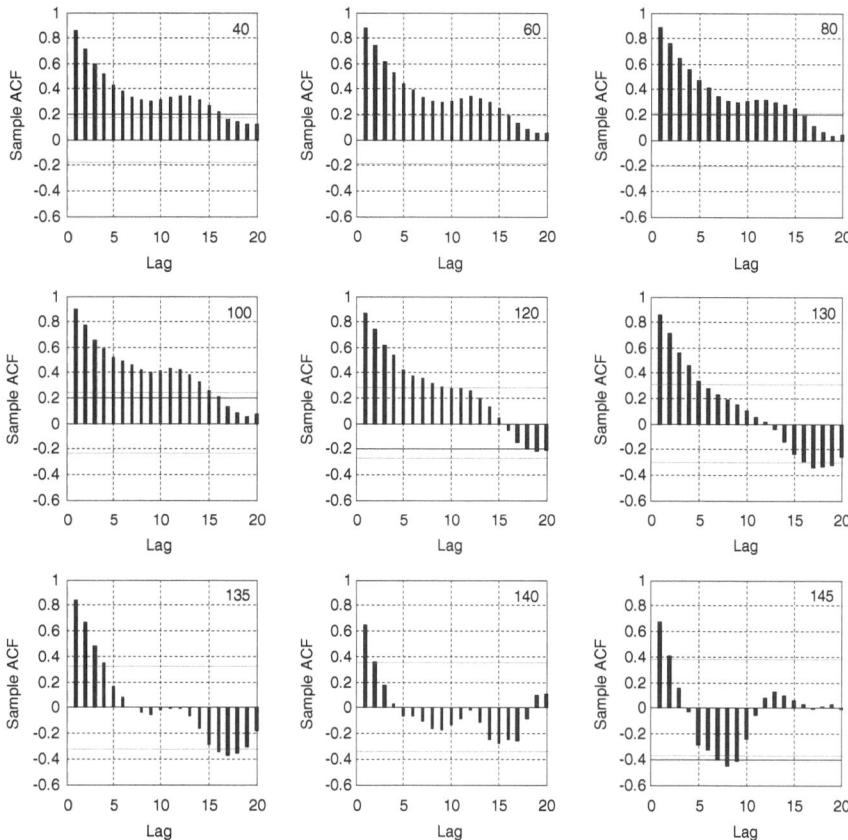

Fig. 4.9 Sample ACF of residuals for differencing lengths of 40, 60, 80 100, 120, 130, 135, 140, 145 years for the period 1837–2009

z_p^- is estimated as -1.75, -1.53, -2.66; and the standardized upper bound z_p^+ is estimated as 1.86, 2.23, 2.15 for ARFIMA models for 1837–1977, 1837–1995, and 1837–2009 periods, respectively. Residuals of the ARFIMA models pass the port-manteau lack of fit test that was applied to the first 15 sample autocorrelations, and the cumulative periodogram test using 5 % Kolmogorov-Smirnov limits.

The list of ARFIMA models, in addition to the ARMA and ARIMA models that were applied to Caspian Sea level time series and their respective RMSE values that were calculated for the forecast validation periods, are tabulated in Table 4.1. ARFIMA model forecasts and their 95 % confidence intervals for the Caspian Sea level after 1977 and 1995 are illustrated in Fig. 4.10. RMSE values are 3.09 m after 1977 and 1.74 m after 1995. These RMSE values are higher than the corresponding values for ARMA and ARIMA models. Although the forecasts of ARMA and ARIMA models after 1977 are better than those of the ARFIMA model, the confidence bands of the ARFIMA model are more realistic. The upper bound of the ARFIMA model forecasts after 1977 is very close to the observed sea level with an RMSE value of 0.12 m. Forecasts and the confidence bands after

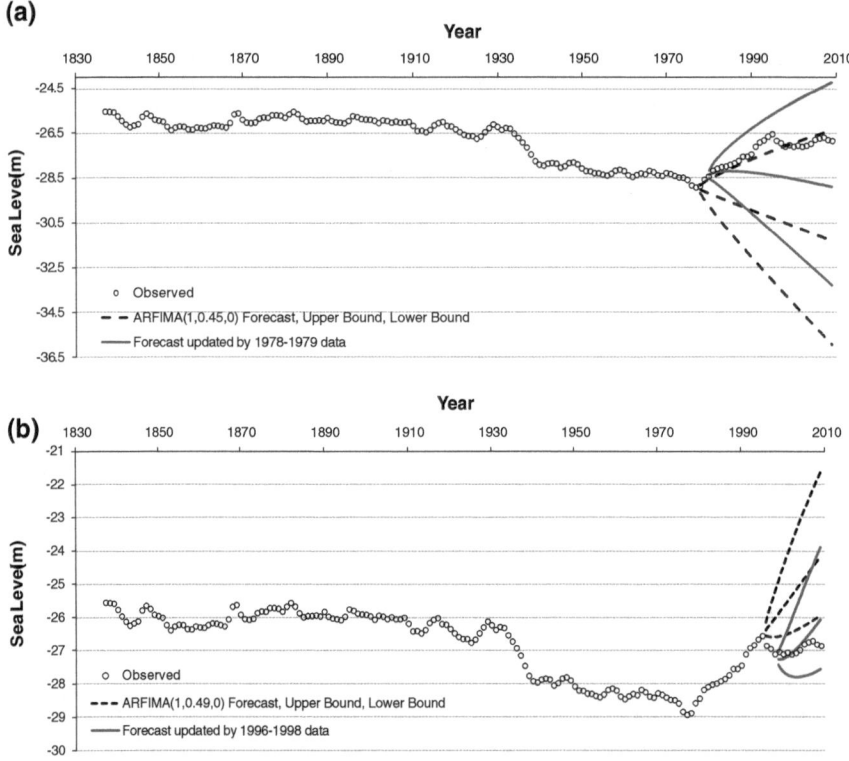

Fig. 4.10 a ARFIMA(1,0.45,0) forecasts after 1977, their 95 % confidence band and the updated forecasts by sea level data of 1978–1979; **b** ARFIMA(1,0.49,0) forecasts after 1995, their 95 % confidence band and the updated forecasts by sea level data of 1996–1998

1977 and 1995 for the ARFIMA models were updated by Eq. (3.11) using the data of 1978–1979, and 1996–1998, respectively. These updated forecasts and their confidence bands are depicted in Fig. 4.10. When the ARFIMA forecasts were updated by the sea level data of 1978–1979 and 1996–1998, the resultant forecasts and their confidence bands improved significantly. RMSE values of the updated ARFIMA forecasts reduce to 1.37 m after 1979, which is less than half of the RMSE value of the forecast after 1977; and to 0.32 m after 1998, which is approximately one-sixth of the RMSE value of the forecast after 1995. After updating, the observed Caspian Sea levels after 1979 are within the confidence bands of the forecasts of the ARFIMA model, as opposed to the confidence bands of the forecasts after 1979 by ARMA and ARIMA models. Furthermore, updating of the forecasts by the ARFIMA model, as shown in Fig. 4.10, has the capability of updating the trends. Hence, the statistical forecasting applications to the Caspian Sea level data show that the ARFIMA model has better capability for adapting to the abrupt sea level changes than the ARMA or ARIMA models.

ARFIMA model forecasts and their 95 % confidence band for the Caspian Sea level after 2009 are illustrated in Fig. 4.11. The Caspian Sea level forecast by the

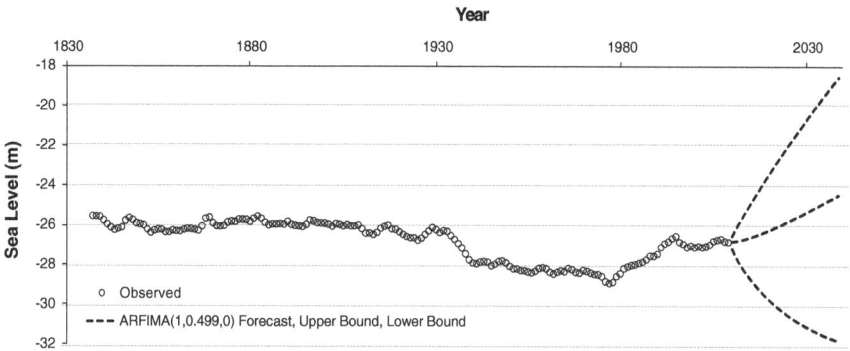

Fig. 4.11 ARFIMA(1,0.499,0) forecast after 2009 and its 95 % confidence band

ARFIMA model in 2039 is −24.5 m; the lower and upper bounds of the 95 % confidence band in 2039 are −31.8 m and −18.6 m. The Caspian Sea level varied between −22 and −35 m in the last 25 centuries according to the paleo-reconstructions, reported by Golitysn (1995). The lower and upper bounds of the 95 % confidence band of the ARFIMA model prediction in 2039 are 3.2 m and 3.4 m above the corresponding paleo-re-constructions reported by Golitysn (1995). Meanwhile, the Caspian Sea level projections in 2039 by seven AOGCMs, presented in Elguindi and Giorgi (2006), range between −31.9 m and −21.6 m with an ensemble mean of −28.8 m for A1B emission scenario and between −32.4 m and −22.1 m with an ensemble mean of −28.1 m for A2 emission scenario. Out of the seven models, only two AOGCMs that use A2 emission scenario and one AOGCM that uses A1b emission scenario predict an increase in the sea level in the twenty-first century. The ARFIMA model predicts an average of 0.081 m/year rise until 2039. The confidence band, estimated by the ARFIMA model, is a measure of the uncertainty involved in forecasting the Caspian Sea level. The size of the confidence band for the ARFIMA model prediction in 2039 is 13.3 m, which is comparable to the level of uncertainty found by Elguindi and Giorgi (2006) from the application of seven AOGCMs (the difference between the minimum and maximum projection in 2039 is approximately 10.3 m for A1b scenario and 10.4 m for A2 scenario). From this comparison of the level of uncertainty of the pure statistical forecasts by the ARFIMA model against the model uncertainty in AOGCMs, it may be inferred that AOGCMs need improvements in their representations of the physical processes that drive Caspian Sea levels.

4.4 Trend Line-ARFIMA Model Forecasts

Between 1837 and 1932, the Caspian Sea level was rather stable, with a mean of −26.05 m and a standard deviation of 0.27 m. The trend line of the observed sea level between 1837 and 1932 (extended until 2009) is shown in Fig. 4.1. The residuals of the observed sea level time series from this linear trend are depicted in Fig. 4.12. Between years 1933 and 1977, the residuals have a mean value of

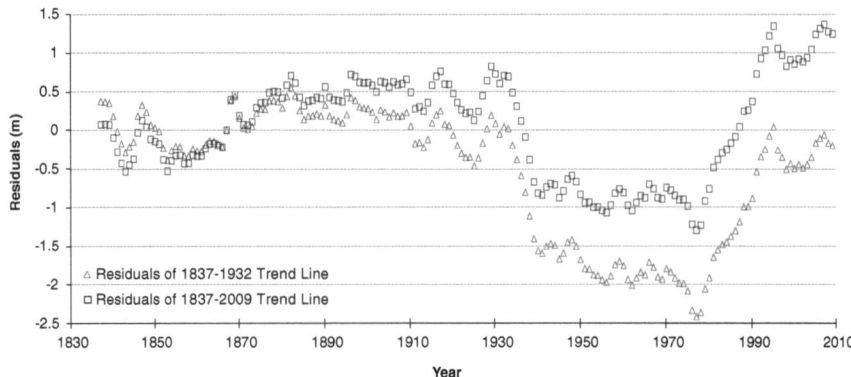

Fig. 4.12   Residuals of 1837–1932 and 1837–2009 trend lines of the Caspian Sea level

−1.62 m, and a minimum value of −2.41 m, observed in 1977. The absolute values of the residuals decrease during the 1978–1995 period, with the mean value of the residuals reducing to −1.19 m. The absolute values of the residuals are less than 0.5 m after 1991. The sudden drop in the Caspian Sea level started in 1932, and the sea level recovery started by the rise in 1977. The sea level trend after 1990s is close to the 1837–1932 trend. The sample ACF, the periodogram, and the logarithm of the periodogram of the Caspian Sea level for the period 1837–2009, after removing the 1837–1932 linear trend, are shown in Fig. 4.13. These three figures in Fig. 4.13 demonstrate that the residuals from the 1837–1932 linear trend still possess long memory. The Hurst numbers H of the original data and the residuals were estimated by the average of the Hurst numbers for the three estimation methods, mentioned earlier. The Hurst numbers before and after the removal of the 1837–1932 trend for 1837–1977, 1837–1995, and 1837–2009 periods are tabulated in Table 4.2. A process has long-range dependence when $0.5 < H < 1$, has short-range dependence when $0 < H < 0.5$ and the observations are uncorrelated when $H = 0.5$. From Table 4.2 it may be inferred that the estimated Hurst numbers indicate long memory in the Caspian Sea level series. For 1837–1977, 1837–1995, and 1837–2009 periods, the fractional difference parameters ($d = H − 0.5$) after removal of the 1837–1932 trend were estimated as 0.43, 0.47, 0.498, respectively, confirming the long memory. Furthermore, the fractional difference parameters, after removal of the 1837–1932 trend, are further away from the stationarity limit ($d = 0.5$) when compared to the corresponding values before the removal of the linear trend.

   ARFIMA models were applied for the 1837–1977, 1837–1995, and 1837–2009 periods to the Caspian Sea level residuals from the 1837–1932 trend line that was extended until 1977, 1995, and 2009, respectively. The models that were developed by combining the Trend Lines with ARFIMA models, are referred to as TL(1837–1932)-ARFIMA(p,d,q) models. The model parameters and the respective RMSE values are tabulated in Table 4.3. The residuals of the combined linear

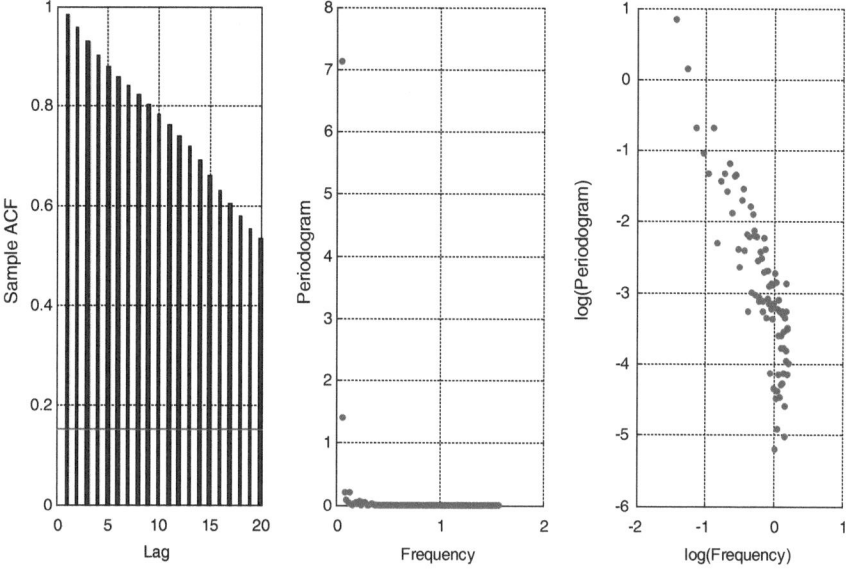

Fig. 4.13 Sample ACF, periodogram, and the logarithm of the periodogram of Caspian Sea levels for the period 1837–2009 after removing the 1837–1932 trend

Table 4.2 The Hurst numbers before and after the removal of the 1837–1932 trend for 1837–1977, 1837–1995, and 1837–2009 periods

Duration	Hurst number	
	Original data	1837–1932 linear trend removed
1837–1977	0.95	0.93
1837–1995	0.99	0.97
1837–2009	0.999	0.998

trend-ARFIMA (TL-ARFIMA) models pass the portmanteau lack of fit test that was applied to the first 15 autocorrelations, and the cumulative periodogram test using 5 % Kolmogorov-Smirnov limits.

TL(1837–1932)-ARFIMA(p,d,q) model forecasts and their 95 % confidence band for the Caspian Sea level after 1977 and 1995 are demonstrated in Fig. 4.14a, b, respectively. RMSE values for these forecasts are 2.52 m after 1977 and 0.46 m after 1995, which are lower than the corresponding values of the ARFIMA-only models (3.09 m and 1.74 m, given in Table 4.1). RMSE value of the TL(1837–1932)-ARFIMA model forecast after 1995 reduced from 1.74 m to 0.46 m because the extended linear trend line, which is part of the TL(1837–1932)-ARFIMA model, represents the observed Caspian sea level trend after 1995. When TL(1837–1932)-ARFIMA forecasts were updated by the Caspian sea level observations during 1978–1979 and 1996–1998 periods, RMSE values of the updated forecasts became 0.99 m after 1979 and 0.50 m after 1998. Updating by observations improved the

Table 4.3 Summary of TL-ARFIMA models that were applied to the Caspian Sea level Time Series

Trend line (duration)	ARFIMA model	Duration	Updated years	Validation period	RMSE
$-0.0042T - 18.206$ (1837–1932)	$(1 - 0.9973B)(1 - B)^{0.43} X_t = \varepsilon_t$	1837–1977	–	1978–2009	2.52
		1837–1977	1978–1979	1980–2009	0.99
$-0.0042T - 18.206$ (1837–1932)	$(1 - 0.8893B)(1 - B)^{0.47} X_t = \varepsilon_t$	1837–1995	–	1996–2009	0.46
		1837–1995	1996–1998	1999–2009	0.50
$-0.0042T - 18.206$ (1837–1932)	$(1 - 0.7886B)(1 - B)^{0.498} X_t = \varepsilon_t$	1837–2009	–	–	–
$-0.0144T + 0.8366$ (1837–2009)	$(1 - 0.9373B)(1 - B)^{0.40} X_t = \varepsilon_t$	1837–2009	–	–	–

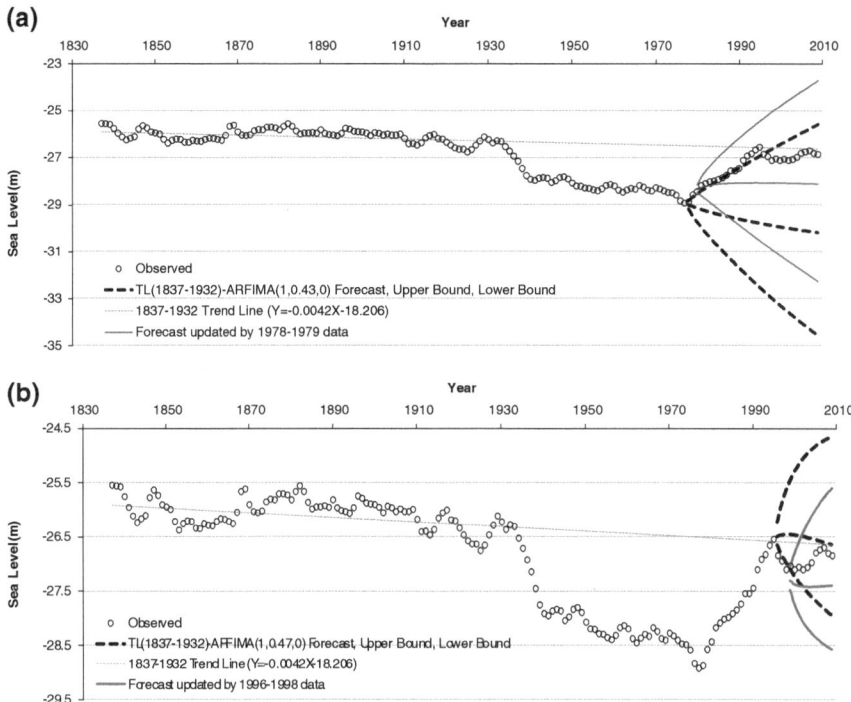

Fig. 4.14 a TL(1837–1932)-ARFIMA(1,0.43,0) combination model forecast after 1977, its 95 % confidence interval and the updated forecast by sea level data of 1978–1979; **b** TL(1837–1932)-ARFIMA(1,0.47,0) forecast after 1995, its 95 % confidence interval and the updated forecast by sea level data of 1996–1998

model forecasts after 1977 but did not improve the forecasts after 1995. Although TL-ARFIMA models may produce better forecasts than ARFIMA-only models if a representative trend line (TL model) can be found, combining the TL model with the ARFIMA model may limit the updating performance of the ARFIMA model.

Finally, the forecasts of the Caspian Sea level after 2009 were investigated for the TL-ARFIMA models. Since the physical reasons for the return of the Caspian sea level time series to their long term trend is not known, the continuation of the 1837–1932 trend is not guaranteed after 2009. Consequently, the 1837–2009 trend was combined with the ARFIMA model to develop a combination model, TL(1837–2009)-ARFIMA(p,d,q). The trend line of the observed sea level between 1837 and 2009 is shown in Fig. 4.1. The residuals of the observed sea level time series from this linear trend are depicted in Fig. 4.12. The mean value of the residuals is 0.05 m and the standard deviation is 0.66 m. TL(1837–2009)-ARFIMA(p,d,q) model parameters are given in Table 4.3.

The forecasts and their 95 % confidence bands for the Caspian Sea level after 2009 for TL(1837–1932)-ARFIMA and TL(1837–2009)-ARFIMA models are depicted in Fig. 4.15a, b, respectively. Unlike the ARFIMA-only model, which

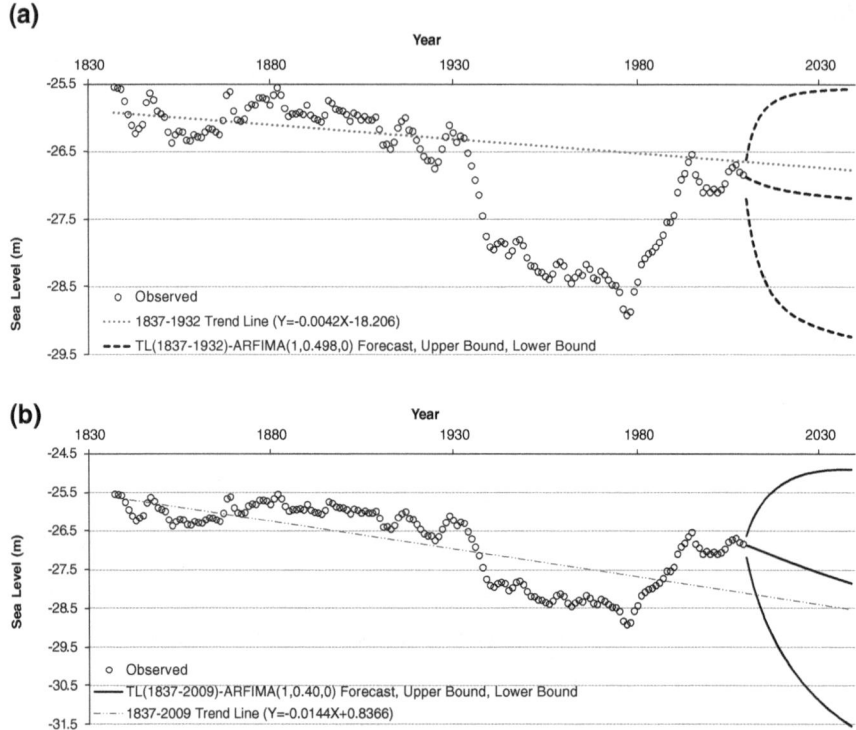

Fig. 4.15 a TL(1837–1932)-ARFIMA(1,0.498,0) and **b** TL(1837–2009)-ARFIMA(1,0.40,0)
models' forecasts after 2009 and their respective 95 % confidence bands

forecasts an average sea level rise of 0.081 m/year until 2039, TL(1837–1932)-
ARFIMA and TL(1837–2009)-ARFIMA models forecast average sea level drops
of 0.011 m/year and 0.033 m/year, respectively. The Caspian Sea level forecast by
TL(1837–1932)-ARFIMA model for 2039 is −27.2 m with the 95 % confidence
interval (−29.2 m, −25.6 m). The TL(1837–2009)-ARFIMA model forecast for
2039 is −27.8 m with the 95 % confidence interval (−31.5 m, −24.9 m). The size
of the forecast 95 % confidence interval in 2039 is the lowest for TL(1837–1932)-
ARFIMA(1,0.498,0) model and is the highest for the ARFIMA(1,0.499,0) model.
As discussed in Sect. 3.2, the forecast confidence interval size depends on the difference parameter d, the variance and probability distribution of the residuals, the
forecast lead time l and the autoregressive and moving average coefficient estimates.

4.5   Conclusions

Forecasting performance of the ARMA, ARIMA, ARFIMA and TL-ARFIMA
models were investigated for the annually averaged Caspian Sea level data, which
are available since 1837. The sample ACFs, periodograms and the estimated Hurst

coefficients demonstrated that the historical Caspian Sea level signal possesses long-range dependence. In order to check the reliability of the developed models, the authors utilized portmanteau lack of fit and cumulative periodogram goodness-of-fit tests as suggested by Box and Jenkins (1976). The ARIMA, ARFIMA and TL-ARFIMA models passed but those of the ARMA models failed the two goodness-of-fit tests.

The following conclusions may be drawn from this study:

i. One of the major challenges in statistical forecasting of a chaotic signal is to predict when to apply what model. The chaotic behavior of signals, like that of Caspian Sea level time series, renders the confidence band estimation and forecast updating components of forecasting quite significant for the forecast performance. In the case of the Caspian Sea level time series the forecast confidence bands and the forecast updating performance of ARFIMA models were shown to be superior compared to those of ARMA or ARIMA models. The updating component of the long memory model, described in Sect. 4.3, renders the forecasting model more reliable, as shown in the Caspian Sea level example. The aim of this study was not to perform point forecasting. Ray (1993) pointed out that the explicit modeling of long range dependence may not be too useful for point forecasting, but it can be important for confidence band estimation.

ii. According to the RMSE values tabulated in Tables 4.1 and 4.3, the forecast performance of ARMA, ARIMA, ARFIMA, and TL-ARFIMA models depends on the time period to which the models are applied. This is true especially for chaotic signals like Caspian Sea level time series.

iii. The lower and upper bounds of the 95 % confidence band (-31.8 m, -18.6 m) for the ARFIMA model sea level forecast for 2039 are 3.2 m and 3.4 m above the corresponding Caspian Sea level limits (as reported by Golitysn 1995) in the last 25 centuries. 13 m Caspian Sea level variation in the last 25 centuries, as reported by Golitysn (1995), is surprisingly very close to the size of the confidence band of the ARFIMA model prediction for 2039, which is 13.3 m. Considering the level of uncertainty in AOGCM forecasts, the pure statistical forecasts as in the case of Caspian Sea level time series reported here, may give valuable insights about the future sea levels without utilizing the computationally intense AOGCM approach.

iv. Similar to the stand-alone ARFIMA forecast (without combination with a long term trend) that was performed in this study, two AOGCMs that used the A2 emission scenario and one AOGCM that used the A1b emission scenario out of seven models that were studied by Elguindi and Giorgi (2006), projected an increase in the Caspian Sea level until 2039. The size of the confidence interval of the ARFIMA model sea level prediction for 2039 is 13.3 m, which is approximately 30 % larger than the difference between the minimum and maximum projections of the AOGCMs for 2039.

v. Although TL-ARFIMA (ARFIMA combined with a long term trend) models may produce better forecasts than ARFIMA-only models if a representative TL model could be found, combining a TL model with the ARFIMA model may limit the updating performance of the ARFIMA model. Furthermore, since the physical reasons for the return of the Caspian Sea level time series to their long term trend is not yet known, the continuation of the 1837–1932 trend is not guaranteed after 2009. The fact that the chaotic dipping of the Caspian Sea level has returned to its long term trend is a fundamental phenomenon that needs physical explanation.

vi. The Caspian Sea level forecast for 2039 is -27.2 m with the 95 % confidence band (29.2 m, -25.6 m) for TL(1837–1932)-ARFIMA model. It is -27.8 m with the 95 % confidence band (-31.5 m, -24.9 m) for TL(1837–2009)-ARFIMA model. TL(1837–1932)-ARFIMA and TL(1837–2009)-ARFIMA models forecast sea level drops averaging 0.011 m/year and 0.033 m/year, respectively. Similarly, five of the seven AOGCMs for A2 and six of the seven AOGCMS for A1b emission scenarios also project drops for the Caspian Sea level, as reported by Elguindi and Giorgi (2006). The size of the 95 % confidence band for the sea level forecast of 2039 is the lowest for TL(1837–1932)-ARFIMA(1,0.498,0) model and is the highest for the ARFIMA(1,0.499,0) model.

vii. While in hydrology various authors have considered long range dependence either by means of stationary long memory models [for example, the fractional Gaussian noise model of Mandelbrot and Van Ness (1968) and Mandelbrot and Wallis (1968)], or by nonstationary time trends [such as in Klemes (1974)], the signal of the Caspian Sea level time series seems to contain both a long term secular trend as well as a long range dependent behavior. As shown in this study, even after removing the long term trend from the Caspian Sea level time series, the residual time series still demonstrate long range dependent behavior. The example of the Caspian Sea level time series demonstrates that both the long range dependence and some secular long term trend may exist together in geophysical phenomena.

References

Barlett MS (1955) Stochastic processes. Cambridge University Press, Cambridge

Beran J (1994) Statistics for long-memory processes. Chapman and Hall, New York

Bobrow SN (1961) The transformation of the Caspian Sea. Soviet Geogr 2:47–64

Box GEP, Jenkins GM (1976) Time series analysis: forecasting and control. Holden-Day, San Francisco

Elguindi N, Giorgi F (2006) Projected changes in the Caspian Sea level for the 21st century based on the latest AOGCM simulations. Geophys Res Lett 33:L08706. doi:10.1029/200 6GL025943

Golitysn G (1995) The Caspian sea level as a problem of diagnosis and prognosis of the regional climate change. Izv Russ Acad Sci Atmos Oceanic Phys 31:366–372

Hald A (1952) Statistical theory with engineering applications. Wiley, New York

Klemes V (1974) The Hurst phenomenon—a puzzle? Water Resour Res 10:675–688

Ljung GM, Box GEP (1978) On a measure of lack of fit in time series models. Biometrika 65:297–303

Mandelbrot BB, Van Ness JW (1968) Fractional Brownian motions, fractional noises and application. Soc Ind Appl Math Rev 10:422–437

Mandelbrot BB, Wallis JR (1968) Noah, Joseph and operational hydrology. Water Resour Res 4:909–920

Ray BK (1993) Modeling long-memory processes for optimal long-range prediction. J Time Ser Anal 14:511–525

Renssen H, Lougheed BC, Aerts JCJH, De Moel H, Ward PJ, Kwadijk JCJ (2007) Simulating long-term Caspian Sea level changes: the impact of Holocene and future climate conditions. Earth Planet Sci Lett 261:685–693

Rodionov SN (1994) Global and regional climate interaction: the Caspian Sea experience. Kluwer, Dordrecht, p 241 pp

Shilo NA (1989) Causes of fluctuations in the level of the Caspian Sea. Trans USSR Acad Sci 305:66–69

Vdovykin GP (1990) On the relationship of changes in Caspian Sea level to neotectonic movements. Trans USSR Acad Sci 310:85–87

Vyushin DI, Kushner PJ (2009) Power-law and long-memory characteristics of the atmospheric general circulation. J Clim 22(11):2890–2904

Chapter 5
Case Study II: Sea Level Change at Peninsular Malaysia and Sabah-Sarawak

Abstract For the region of Peninsular Malaysia and Malaysia's Sabah-Sarawak northern region of Borneo Island, long sea level records do not exist. In such case the Atmospheric-Oceanic Global Climate Model (AOGCM) projections for the 21st century can be downscaled to the Malaysia region by means of regression techniques, utilizing the short records of satellite altimeters in this region against the GCM projections during a mutual observation period. In this case study on the assessment of sea level change along the coastlines of Peninsular Malaysia and Sabah-Sarawak, the spatial variation of the sea level change is estimated in time by assimilating the global mean sea level projections from the AOGCM simulations to the satellite altimeter observations along the subject coastlines. Details of this case study were presented in Ercan et al. (2013) at Hydrol Process, 27(3):367–377.

Keywords Rregression techniques, • Satellite altimeter observations • GCM projections • Global mean sea level

5.1 Introduction

For the region of Peninsular Malaysia and Malaysia's Sabah-Sarawak northern region of Borneo Island, long sea level records do not exist. In such case the Atmospheric-Oceanic Global Climate Model (AOGCM) projections for the twenty-first century can be downscaled to the Malaysia region by means of regression techniques, utilizing the short records of satellite altimeters in this region against the GCM projections during a mutual observation period. In this case study on the assessment of sea level change along the coastlines of Peninsular Malaysia and Sabah-Sarawak, the spatial variation of the sea level change is estimated in time by assimilating the global mean sea level projections from the AOGCM simulations to the satellite altimeter observations along the subject coastlines. Details of this case study were presented in Ercan et al. (2013).

A. Ercan et al., *Long-Range Dependence and Sea Level Forecasting*,
SpringerBriefs in Statistics, DOI: 10.1007/978-3-319-01505-7_5, © The Author(s) 2013

Climate models provide credible quantitative estimates of future climate change, particularly at continental scales and above (Randall et al. 2007). However, due to their coarse spatial grid resolution, their description of the spatial variation of the sea level change at regional and smaller spatial scales is too coarse. In analyzing Fig. 10.32 in Meehl et al. (2007a) for the projected geographical variation of local sea level change, if one compares this spatial variation with the observed spatial distributions of the sea level change in Figs. 5.15a and 5.16a in Bindoff et al. (2007), one may conclude that the projected spatial distribution of sea level change by AOGCMs does not account for the observed spatial distribution, especially over the Southeast Asia region. Therefore, a prudent projection in a region where the local ground-based observations are for a short time period, could use the AOGCM projections for the global average sea level change, but then distribute these projections in space over the Peninsular Malaysia and Sabah-Sarawak coastlines according to the observed patterns based on the satellite altimetry data. Hence, in this study the spatial distribution of the sea level change along the Peninsular Malaysia and Sabah-Sarawak coastlines that was observed by satellite altimeters in time, is merged with the AOGCM projections of the global mean sea level change during the twenty-first century in order to better predict the spatial variation of the local sea level change along the coastal regions of Malaysia.

The confidence in the predictions of AOGCM simulations comes from the founding of the models in accepted physical principles and from their ability t reproduce observed features of current climate and past climate evolution (Randall et al. 2007). In this study the global mean sea level estimates in monthly intervals from various AOGCM simulations under climate scenarios 20C3 M (the scenario representing the climate of the twentieth Century), and three SRES greenhouse emission scenarios (B1, A1B and A2) were obtained from the World Climate Research Programme's (WCRP's) Coupled Model Intercomparison Project phase 3 (CMIP3) multi-model dataset (Meehl et al. 2007b). The number of AOGCM simulations that were performed by each model for each of the four scenarios is tabulated in Table 5.1. A total

Table 5.1 Number of AOGCM simulations (models and applied emission scenarios) used to assess sea level rise around Peninsular Malaysia and Sabah-Sarawak coastlines Ercan et al. (2013)

Model	Number of simulations per scenario			
	20C3 M	SRES B1	SRES A2	SRES A1B
CGCM3.1 (2004)[a]	1	1	1	1
GISS-AOM (2004)[b]	2	2	0	2
GISS-ER (2004)[c]	9	1	1	5
MIROC3.2(hires) (2004)[d]	1	1	0	1
MIROC3.2(medres) (2004)[d]	3	3	3	3
ECHO-G (1999)[e]	3	3	3	3
MRI-CGCM2.3.2a (2001)[f]	5	5	5	5

[a]McFarlane et al. (1992), Flato (2005), Pacanowski et al. (1993)
[b]Russell et al. (1995), Russell (2005)
[c]Schmidt et al. (2006), Russell et al. (1995)
[d]K-1 Developers (2004)
[e]Roeckner et al. (1996), Legutke and Maier-Reimer (1999)
[f]Shibata et al. (1999), Yukimoto et al. (2001)

of 73 projections were analyzed for the seven AOGCMs available: 24 projections for the 20C3 M scenario, 16 projections for the SRES B1 scenario, 13 projections for the SRES A2 scenario, and 20 projections for the SRES A1B scenario. References to the AOGCMs used were reported in Ercan et al. (2013).

5.2 Observed Satellite Altimeter Data

In this case study, linear regression analyses were performed both for monthly tidal gauge and monthly satellite altimeter observations along Peninsular Malaysia and Sabah-Sarawak coastlines. The results of these analyses showed that slopes of the linear trend lines are significantly greater for the satellite altimeter observations when compared to those for the tidal gauge observations. Meanwhile, there is no missing data in satellite altimeter observations for any month during the observation period, and the uncertainties in satellite altimeter observations are well described. Furthermore, it was possible to correct the errors in the satellite observations. Therefore, the satellite altimeter data were utilized as the basis for assimilating the future sea level projections that are derived from the global mean sea level projections from the AOGCM simulations during the twenty-first century, to locations around Peninsular Malaysia and Sabah-Sarawak coastlines.

The combined TOPEX/Poseidon, Jason-1 and Jason-2/OSTM sea level fields in monthly intervals (CSIRO 2010) were used in the satellite altimeter data linear trend analyses. For the satellite altimeter data, the annual and semi-annual signals were removed; and the inverse barometer and glacial isostatic adjustment (GIA) corrections were done (CSIRO 2010). In order to perform a sensitivity analysis for these corrections, the satellite altimeter data with and without these corrections were analyzed. The sea level rise rates that were calculated by linear regression analyses of the satellite altimeter data around Peninsular Malaysia coastline with and without the application of the three correction methods are tabulated in Table 5.2.

Around Peninsular Malaysia coastline, the sea level rise rates that were calculated from the satellite altimeter data when the annual and semi-annual signals were removed, are less by an average of 0.29 mm/year for Peninsular Malaysia when compared to the rates without removing the signals. The Inverse Barometer is the correction for variations in the sea surface height due to atmospheric pressure variations (Ponte and Gaspar 1999; Dorandeu and Le Traon 1999). Around Peninsular Malaysia coastline, the sea level rise rates that were calculated from the satellite altimeter data with the inverse barometer correction are smaller with an average of 0.43 mm/year for Peninsular Malaysia when compared to the rates without any correction. Modern measurements of the rate of sea level rise are significantly contaminated by the influence of the ongoing process of Glacial Isostatic Adjustment (GIA) due to the most recent deglaciation event of the Late Quaternary ice-age (Peltier 2009). The GIA correction applied to Topex/Poseidon-derived altimetric measurements was demonstrated in Peltier (2002). Using the ICE-4G (VM2) model of the GIA process, described in Peltier (1994, 1996), analyses demonstrated that such measurements would be biased down by approximately

Table 5.2 Sea level rise rates (mm/yr) calculated by linear regression analyses of the satellite altimeter data around Peninsular Malaysia coastline with and without applying the correction methods Ercan et al. (2013)

Longitude/Latitude	Case 1[a]	Case 2[b]	Case 3[c]	Case 4[d]	Case 5[e]
Peninsular Malaysia					
100E/6 N	6.33	6.63	6.21	5.94	6.08
99E/5 N	6.47	6.82	6.53	6.11	6.45
104E/1 N	4.49	4.90	4.17	3.95	3.87
105E/2 N	4.23	4.65	3.91	3.70	3.68
104E/3 N	3.37	3.73	3.03	2.94	2.88
104E/4 N	3.22	3.59	2.87	2.79	2.73
104E/5 N	3.26	3.58	2.85	2.90	2.78
103E/6 N	4.04	4.34	3.59	3.59	3.46
103E/7 N	4.15	4.46	3.67	3.68	3.49
102E/7 N	4.94	5.23	4.49	4.46	4.29
101E/7 N	5.80	6.05	5.38	5.33	5.20
99E/6 N	5.75	6.05	5.73	5.46	5.70
99E/7 N	5.15	5.40	5.04	4.82	5.02
Average	4.71	5.03	4.42	4.28	4.28
GMSL[f]	2.94	3.38	2.92	2.81	3.22

[a]No correction
[b]GIA correction was performed
[c]semi-annual signals were removed
[d]Inverse barometer correction was performed
[e]Annual and semi-annual signals were removed; Inverse barometer and GIA corrections were performed
[f]Global Mean Sea Level

0.3 mm/year. Table 5.2 depicts that the sea level rise rates that were calculated from the satellite altimeter data with GIA correction are greater with an average of 0.32 mm/year for Peninsular Malaysia than the rates without any correction. The sea level rise rates that were calculated from the satellite altimeter data, when all the three corrections are applied, are smaller with an average of 0.43 mm/year for Peninsular Malaysia than the rates without any correction. When all the three corrections are applied, the averages of the sea level rise rates around Peninsular Malaysia coastline that were calculated from the satellite altimeter data between January 1993 and March 2010 are 1.06 mm/year larger than the global average.

5.3 Assimilating AOGCM Simulations to Satellite Observations

The determination of the variation of the sea level change with respect to the spatial location along the Peninsular Malaysia and Sabah-Sarawak coastlines is based on the linear trend analyses of the observed satellite altimetry data. Using monthly satellite altimeter data of January 1993–December 2000 and using monthly twentieth century global mean sea level predictions of various AOGCMs, one can write the below regression equation at each satellite altimeter location i and for AOGCM j, and estimate the equation coefficients a_{ij} and b_{ij} such that $\sum_{k} \left(\varepsilon_{i,j,k} \right)^2$ is minimized for each altimeter location and AOGCM (Ercan et al. 2013).

$$y_{SA,i,j,k} = a_{i,j} y_{GMSL\_20c3m,j,k} + b_{i,j} + \varepsilon_{i,j,k} \tag{5.1}$$

Here y_{SA} is the satellite observation of the sea level, y_{GMSL_20c3m} is 20th century global mean sea level (GMSL) prediction, and ε is the error term. Then one can estimate the sea level y at satellite altimeter location i at time k using twenty-first century global mean sea level projections of AOGCM j for SRES B1, A1B and A2 scenarios by (Ercan et al. 2013)

$$y_{sresb1,i,j,k} = a_{i,j} y_{GMSL\_sresb1,j,k} + b_{i,j} \tag{5.2}$$

$$y_{srea1b,i,j,k} = a_{i,j} y_{GMSL\_srea1b,j,k} + b_{i,j} \tag{5.3}$$

$$y_{sresa2,i,j,k} = a_{i,j} y_{GMSL\_sresa2,j,k} + b_{i,j} \tag{5.4}$$

where a_{ij} and b_{ij} are linear regression coefficients estimated from Eq. (5.1) for satellite altimeter location i and AOGCM j. After solving Eqs. (5.2)–(5.4) for each satellite altimeter location i and for each AOGCM j at time k in the twenty-first century, means and 95 percent confidence bands of the sea level rise rates and corresponding sea level rise estimates could be obtained for twenty-year intervals in the twenty-first century using the three emission scenarios. The linear regressions between the observed local sea level and global mean sea level in the late twentieth century, simulated by AOGCMs, are assumed to hold during the twenty-first century. However, the nonlinearity in AOGCM projections are reflected in the local sea level projections through the regression relationships that are established by means of the historical data.

5.4 Sea Level Change

Means and lower and upper bounds of 95 percent confidence intervals of the sea level change rate projections around Peninsular Malaysia and Sabah-Sarawak coastlines are depicted in Figs. 5.1, 5.2, 5.3, respectively using the ensemble of all the three SRES scenarios and all the available AOGCMs (tabulated in Table 5.1). These figures show the average sea level change rates in twenty-year increments in the twenty-first century and the average rate during 2001–2100. The highest sea level rise rate occurs at 100E/6 N location with mean 5.17 mm/yr with a confidence interval of (0.02, 23.03) mm/yr at Peninsular Malaysia and at 119E/4 N location with mean 10.64 mm/yr with a confidence interval of (0.00, 43.86) mm/yr at Sabah and Sarawak during 2001–2100. The lowest sea

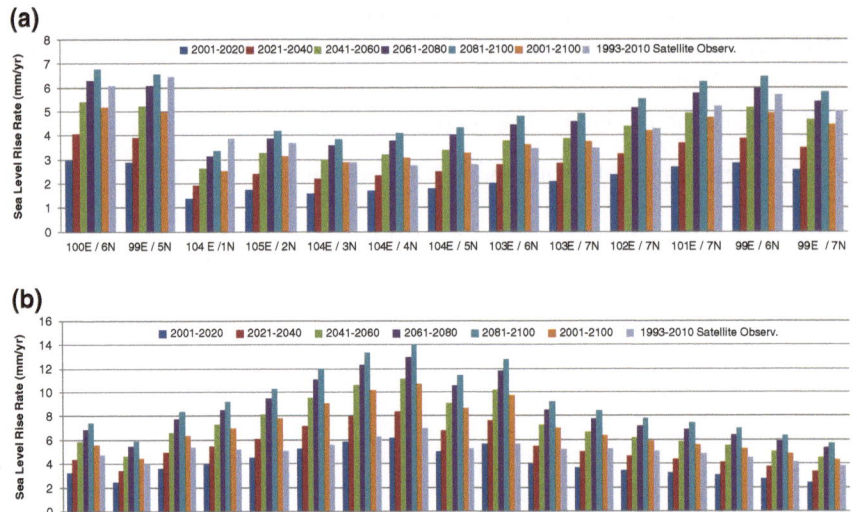

Fig. 5.1 Mean sea level rise rate (mm/yr) projections by means of the assimilated available AOGCM projections using the ensemble of SRES B1, A1B and A2 scenarios in the twenty-first century: **a** around Peninsular Malaysia coastline, **b** around Sabah and Sarawak coastline Ercan et al. (2013)

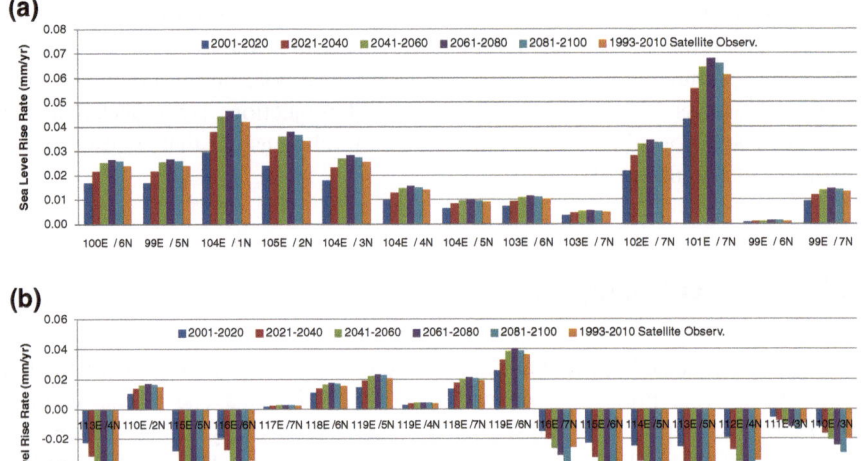

Fig. 5.2 Lower bounds of 95 percent confidence interval of sea level rise rate (mm/yr) projections by means of the assimilated available AOGCM projections using the ensemble of SRES B1, A1B and A2 scenarios in the twenty-first century: **a** around Peninsular Malaysia coastline, **b** around Sabah and Sarawak coastline Ercan et al. (2013)

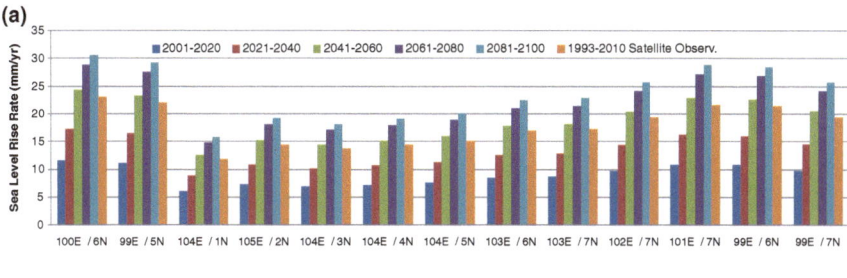

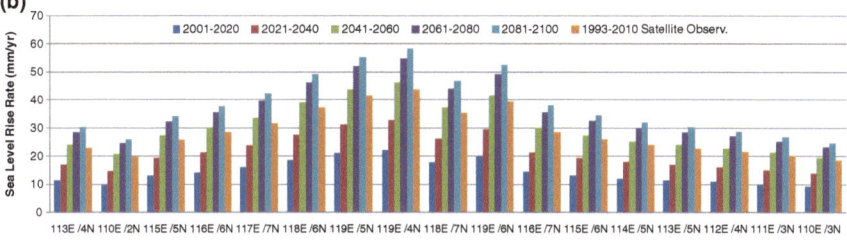

Fig. 5.3 Upper bounds of 95 percent confidence interval of sea level rise rate (mm/yr) projections by means of the assimilated available AOGCM projections using the ensemble of SRES B1, A1B and A2 scenarios in the twenty-first century: **a** around Peninsular Malaysia coastline, **b** around Sabah and Sarawak coastline Ercan et al. (2013)

Table 5.3 Summary of mean sea level rise rate (mm/yr) predictions and 95 percent confidence intervals from all available AOGCM projections around Peninsular Malaysia and Sabah and Sarawak coastlines for the ensemble of SRES B1, A1B and A2 scenarios (LB: lower bound; UB: upper bound)

	Time Interval					
	2001 2020	2021–2040	2041–2060	2061–2080	2081–2100	2001–2100
Peninsular Malaysia						
LB	0.02	0.02	0.02	0.02	0.02	0.02
Mean	2.19	3.02	4.06	4.77	5.15	3.90
UB	8.92	13.24	18.67	22.15	23.54	17.73
Sabah-Sarawak						
LB	−0.01	−0.01	−0.01	−0.02	−0.02	−0.01
Mean	4.02	5.47	7.29	8.51	9.21	6.98
UB	14.45	21.43	30.23	35.86	38.11	28.70

level rise rate occurs at 104E/1 N location with mean 2.53 mm/yr with a confidence interval of (0.04, 11.89) mm/yr at Peninsular Malaysia and at 110E/3 N location with mean 4.32 mm/yr with a confidence interval of (-0.02, 18.52) mm/yr at Sabah and Sarawak during 2001-2100. In the twenty-first century, Figs. 5.1-5.3 show clearly that the means and upper bounds of 95 percent confidence intervals of sea level rise rates are increasing with time toward the future both for Peninsular Malaysia and for Sabah and Sarawak coastlines.

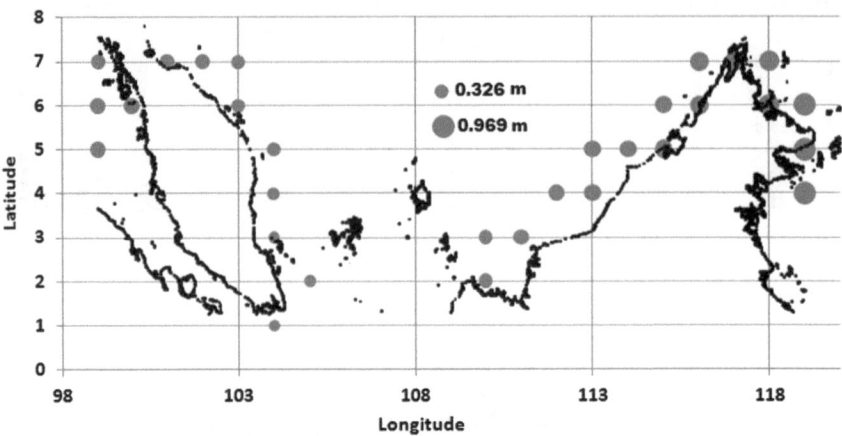

Fig. 5.4 Means of the sea level rise projections around Malaysia coastlines using the ensemble of SRES B1, A1B and A2 scenarios in 2100 Ercan et al. (2013)

Summary of mean sea level rise rate projections in mm/yr with the corresponding 95 percent confidence intervals from all available AOGCM projections around Peninsular Malaysia and Sabah and Sarawak coastlines for the ensemble of SRES B1, A1B and A2 scenarios are tabulated in Table 5.3. When all the three SRES scenarios for the whole coast of Peninsular Malaysia during 2001–2100 are considered, the mean sea level rise rate is 3.90 mm/year with the confidence interval of (0.02, 17.73) mm/yr. On the other hand, when all the three SRES scenarios for the whole coast of Sabah and Sarawak during 2001–2100 are considered, the mean sea level rise rate is 6.98 mm/year with the confidence interval of (−0.01, 28.70) mm/yr.

The sea level rise can be estimated by multiplying the sea level rise rate estimate by the duration. Means of the sea level rise projections in 2100 (since year 2000) are depicted on the map of Peninsular Malaysia and Sabah-Sarawak in Fig. 5.4, where sea level rises are represented by the size of the circles.

In 2100, the highest sea level rise occurs at 100E/6 N location with mean 0.517 m with a confidence interval of (0.002, 2.303) m at Peninsular Malaysia; and at 119E/4 N location with mean 1.064 m with a confidence interval of (0.000, 4.386) m at Sabah and Sarawak. In 2100, the lowest sea level rise occurs at 104E/1 N with mean 0.253 m with a confidence interval of (0.004, 1.189) m at Peninsular Malaysia; and at 110E/3 N location with mean 0.432 m with a confidence interval of (−0.002, 1.852) m at Sabah and Sarawak.

5.5 Conclusions

In this case study on the assessment of sea level change along the coastlines of Peninsular Malaysia and Sabah-Sarawak, the spatial variation of the sea level change was estimated by assimilating the global mean sea level rise projections from the AOGCM simulations to the satellite altimeter observations along these coastlines.

The sea level around the Peninsular Malaysia coastlines is projected by means of the assimilated AOGCM projections to rise with a mean between 0.253 and 0.517 m in 2100. The upper bound of the 95 percent confidence interval for the sea level rise around Peninsular Malaysia is between 1.189 and 2.303 m in 2100. The highest sea level rise occurs at the north-east and north-west regions of Peninsular Malaysia. The sea level rise estimates that are based solely on the local observations by satellite altimeters around Peninsular Malaysia are between 0.273 and 0.645 m in 2100 (assuming the observed rate continues in the twenty-first century). These estimates are close to the mean projections that are assimilated from the AOGCM projections to the Peninsular Malaysia coastal areas by means of the local observations, and are within the 95 percent confidence intervals of the mean projections.

The sea level around Sabah and Sarawak coastlines is projected by means of the assimilated AOGCM projections to rise with a mean between 0.432 and 1.064 m in 2100. The upper bound of the 95 percent confidence interval for the sea level rise at Sabah and Sarawak is between 1.852 and 4.386 m in 2100. The highest sea level rise at Sabah and Sarawak is estimated to occur at north and east sectors of Sabah. The sea level rise estimates that are based solely on the local observations by satellite altimeters around Sabah and Sarawak are between 0.382 and 0.700 m in 2100. These estimates are close to the mean projections that are assimilated from the AOGCM projections to Sabah and Sarawak coastal areas by means of the local observations, and are within the 95 percent confidence intervals of the mean projections.

Elevation maps of the coastal regions of the study area with high horizontal grid resolution and high vertical accuracy are necessary for the performance of realistic sea inundation analyses. The vertical accuracy should be at least in the order of centimeters in order to capture the spatial details of the sea level rise. While it is difficult to specify a specific value for the horizontal grid resolution, it should capture local high and low areas.

References

Bindoff NL, Willebrand J, Artale V, Cazenave A, Gregory J, Gulev S, Hanawa K, Le Quéré C, Levitus S, Nojiri Y, Shum CK, Talley LD, Unnikrishnan A (2007) Observations: oceanic climate change and sea level. In: Solomon S et al. (eds) Climate change 2007: The Physical Science Basis. Contribution of working group I to the fourth assessment report of the intergovernmental panel on climate. Cambridge University Press: Cambridge, New York, pp 385–432

Commonwealth Scientific and Industrial Research Organisation (CSIRO) (2010) http://www.cmar.csiro.au/sealevel/sl_data_cmar.html. Downloaded in Sept. 2010

Dorandeu J, Le Traon PY (1999) Effects of global mean atmospheric pressure variations on mean sea level changes from TOPEX/Poseidon. J Atmos Oceanic Technol 16(9):1279–1283

Ercan A, Mohamad MF, Kavvas ML (2013) Sea level rise due to climate change around the Peninsular Malaysia and Sabah and Sarawak coastlines for the 21st century. Hydrol Process 27(3):367–377. doi:10.1002/hyp.9232

Flato GM (2005) The Third Generation Coupled Global Climate Model (CGCM3) (and included links to the description of the AGCM3atmospheric model). http://www.cccma.bc.ec.gc.ca/models/cgcm3.shtml

K-1 model developers (2004) K-1 coupled model (MIROC) description. Technical report 1. Center for Climate System Research, University of Tokyo

Legutke S, Maier-Reimer E (1999) Climatology of the HOPE-G Global Ocean General Circulation Model. Technical report No. 21, German Climate Computer Centre (DKRZ): Hamburg, Germany, pp 90

Meehl GA, Stocker TF, Collins WD, Friedlingstein P, Gaye AT, Gregory JM, Kitoh A, Knutti R, Murphy JM, Noda A, Raper SCB, Watterson IG, Weaver AJ, Zhao Z-C (2007a) Global climate projections. In: Solomon S et al. (eds) Climate change 2007: The Physical Science Basis. Contribution of working group I to the fourth assessment report of the intergovernmental panel on climate change. Cambridge University Press: Cambridge, New York

Meehl GA, Covey C, Delworth T, Latif M, McAvaney B, Mitchell JFB, Stouffer RJ, Taylor KE (2007b) The WCRP CMIP3 multimodel dataset: a new era in climate change research. Bull Am Meteorol Soc 88:1383–1394

McFarlane NA, Boer GJ, Blanchet J-P, Lazare M (1992) The Canadian Climate Centre second-generation general circulation model and its equilibrium climate. J Climate 5(10):1013–1044.

Pacanowski RC, Dixon K, Rosati A (1993) The GFDL Modular Ocean Model Users Guide, Version 1.0. GFDL Ocean Group Technical Report No. 2, Geophysical Fluid Dynamics Laboratory: Princeton, NJ

Peltier WR (1994) Ice-age paleotopography. Science 265:195–201

Peltier WR (1996) Mantle viscosity and ice-age ice-sheet topography. Science 273:1359–1364

Peltier WR (2002) Global glacial isostatic adjustment: paleo-geodetic and space geodetic tests of the ICE-4G(VM2) model. J Quat Sci 17:491–510

Peltier WR (2009) Closure of the budget of global sea level rise over the GRACE era: the importance and magnitudes of the required corrections for the influence of global glacial isostatic adjustment. Quat Sci Rev 28:1658–1674

Ponte RM, Gaspar P (1999) Regional analysis of the inverted barometer effect over the global ocean using Topex/Poseidon data and model results. J Geophys Res, 104(C7): 15587–15601

Roeckner E, Arpe K, Bengtsson L, Christoph M, Claussen M, Dümenil L, Esch M, Giorgetta M, Schlese U, Schulzweida U (1996) The Atmospheric General Circulation Model ECHAM4: Model Description and Simulation of Present-Day Climate. MPI Report No. 218, Max-Planck-Institut für Meteorologie: Hamburg, Germany, pp 90

Russell GL, Miller JR, Rind D (1995) A coupled atmosphere–ocean model for transient climate change studies. Atmos.-Ocean 33(4):683–730

Russell GL (2005) 4x3 atmosphere–ocean model documentation. http://aom.giss.nasa.gov/doc4x3.html

Randall DA, Wood RA, Bony S, Colman R, Fichefet T, Fyfe J, Kattsov V, Pitman A, Shukla J, Srinivasan J, Stouffer RJ, Sumi A, Taylor KE (2007) Cilmate models and their evaluation. In: Solomon S et al. (eds) Climate change 2007: The Physical Science Basis. Contribution of working group I to the fourth assessment report of the intergovernmental panel on climate change. Cambridge University Press: Cambridge, New York

Schmidt GA, Ruedy R, Hansen JE, et al. (2006) Present day atmospheric simulations using GISS ModelE: Comparison to in-situ, satellite and reanalysis data. Journal of Climate 19(2):153–192.

Shibata K, Yoshimura H, Ohizumi M, Hosaka M, Sugi M (1999) A simulation of troposphere, stratosphere and mesosphere with an MRI/JMA98 GCM. Papers in Meteorology and Geophysics 50(1):15–53

Yukimoto S, Noda A, Kitoh A, Sugi M, et al. (2001) The new Meteorological Research Institute global ocean–atmosphere coupled GCM (MRI-CGCM2)-Model climate and variability. Papers in Meteorology and Geophysics 51(2):47–88

Chapter 6
Summary and Conclusion

Abstract Determining when to apply what model is a major challenge in statistical forecasting of a chaotic signal. The chaotic behavior of signals, like that of Caspian Sea level time series, renders the confidence band estimation and forecast updating components of forecasting quite significant for the forecast performance. In this chapter, a brief summary and conclusions are provided for the monograph "Long-Range Dependence and Sea Level Forecasting".

Keywords Llong-range dependence • Sea level forecasting • Chaotic signals • ARFIMA models

Determining when to apply what model is a major challenge in statistical forecasting of a chaotic signal. The chaotic behavior of signals, like that of Caspian Sea level time series, renders the confidence band estimation and forecast updating components of forecasting quite significant for the forecast performance. The long-range dependence concept, and the methodologies for the estimation of long-range dependence index (Hurst Number) were presented in Chap. 2. The forecasting methodology for ARFIMA models, the uncertainty estimation of forecasts and the updating as new data become available were provided in Chap. 3. It was shown that the forecast confidence interval size depends on the probability distribution of the residuals, forecast lead time, the difference parameter d, and the autoregressive and the moving average coefficients for ARFIMA models.

In Chap. 4, the forecasting performance of the ARMA, ARIMA, ARFIMA and TL-ARFIMA models were investigated for the annually averaged Caspian Sea level data, which are available since 1837. The forecast confidence bands and the forecast updating performance of ARFIMA models were shown to be superior compared to those of ARMA or ARIMA models. The updating component of the long memory model makes the forecasting model more reliable as shown in the Caspian Sea level example. Considering the level of uncertainty in AOGCM forecasts, the pure statistical forecasts such as for the Caspian Sea level case reported

A. Ercan et al., *Long-Range Dependence and Sea Level Forecasting*,   49
SpringerBriefs in Statistics, DOI: 10.1007/978-3-319-01505-7_6, © The Author(s) 2013

here may give valuable insights about the future sea levels without utilizing the computationally intense AOGCM approach.

While in hydrology various authors have considered long range dependence either by means of stationary long memory models [for example, the fractional Gaussian noise model of Mandelbrot and Van Ness (1968) and Mandelbrot and Wallis (1968)], or by nonstationary time trends (such as in Klemes (1974)), the signal of the Caspian Sea level time series seems to contain both a long term secular trend as well as long range dependent behavior. The example of the Caspian Sea level time series has shown that both the long range dependence and some secular long term trend may exist together in geophysical phenomena, and statistical modeling of a time series may be performed by the combination of a trend component and a long memory component.

Instead of the infinitely long differencing lengths, finite differencing lengths for the ARFIMA models were utilized due to the finite duration of the available sea level record. Sample ACFs of the residuals were compared for various differencing lengths, and the one that minimizes the correlation structure in the sample ACFs was selected. Confidence intervals and the forecast updating methodology, provided for ARIMA models in Box and Jenkins (1976), were modified for the ARFIMA models. The confidence intervals of the forecasts were estimated utilizing the probability densities of the residuals without assuming a known distribution. In the literature, normal distribution of the residuals is usually assumed for the estimation of the confidence intervals.

In order to check the statistical model reliability, portmanteau lack of fit and cumulative periodogram tests as model diagnostic tools (Box and Jenkins 1976) were introduced and utilized in the Caspian Sea level example case.

Sea level change has been also studied by AOGCMs (Gregory et al. 2001; Meehl et al. 2007a). There is substantial variability and uncertainty in the spatial distribution of sea level change among all GCMs (Meehl et al. 2007a). Climate models provide credible quantitative estimates of future climate change, particularly at continental scales and above (Randall et al. 2007). However, due to their coarse spatial grid resolution, their description of the spatial variation of the sea level change at regional and smaller spatial scales is too coarse.

In the case study of Peninsular Malaysia and Sabah-Sarawak coastlines on the assessment of sea level change along the coastlines of Peninsular Malaysia and Sabah-Sarawak, as reported in Chap. 5, the spatial variation of the sea level change was estimated by assimilating the global mean sea level projections from the AOGCM simulations to the satellite altimeter observations along the Malaysian coastlines (Ercan et al. 2013). The determination of the variation of the sea level change with respect to the spatial location along the Peninsular Malaysia and Sabah-Sarawak coastlines was based on the linear trend analyses of the observed satellite altimetry data. Using the observed monthly satellite altimeter data and using monthly twentieth century global mean sea level projections of various AOGCM models, a regression equation at each satellite altimeter location for each AOGCM was written, and the corresponding regression coefficients were estimated. The highest sea level rise occurs at the north-east and north-west regions of Peninsular Malaysia and at north and east sectors of Sabah.

In the future, sea inundation studies with fine resolution topographic maps can be performed based on the sea level projections of Caspian Sea and of the coastlines of Malaysia with priority given to urban, industrial, agricultural, touristic, and historical areas. Based on these projections, the impact of the sea level change in these regions can then be evaluated.

Statistical modeling and forecasting approaches may be investigated for other geophysical time series which may also exhibit a trend and a long memory component, as was found in the historical mean sea level data of the Caspian Sea levels.

References

Box GEP, Jenkins GM (1976) Time series analysis: forecasting and control. Holden-Day, San Fransisco

Ercan A, Mohamad MF, Kavvas ML (2013) Sea level rise due to climate change around the Peninsular Malaysia and Sabah and Sarawak coastlines for the 21st century. Hydrol Process 27(3):367–377. doi: 10.1002/hyp.9232

Gregory JM, Church JA, Boer GJ, Dixon KW, Flato GM, Jackett DR, Lowe JA, O'Farrell SP, Roeckner E, Russell GL, Stouffer RJ, Winton M (2001) Comparison of results from several AOGCMs for global and regional sea-level change 1900–2100. Clim Dyn 18(3–4):225–240

Klemes V (1974) The Hurst phenomenon: a puzzle? Water Resour Res 10:675–688

Mandelbrot BB, Wallis JR (1968) Noah, Joseph and operational hydrology. Water Resour Res4:909–920

Mandelbrot BB, Van Ness JW (1968) Fractional Brownian motions, fractional noises and application, Soc. Ind. Appl. Math. Rev., 10, 422–437

Meehl GA, Stocker TF, Collins WD, Friedlingstein P, Gaye AT, Gregory JM, Kitoh A, Knutti R, Murphy JM, Noda A, Raper SCB, Watterson IG, Weaver AJ, Zhao ZC (2007a) Global climate projections. In: Solomon S et al. (eds.) Climate change 2007: the physical science basis. Contribution of working group I to the fourth assessment report of the intergovernmental panel on climate change. Cambridge University Press, Cambridge

Randall DA, Wood RA, Bony S, Colman R, Fichefet T, Fyfe J, Kattsov V, Pitman A, Shukla J, Srinivasan J, Stouffer RJ, Sumi A, Taylor KE (2007) Cilmate models and their evaluation. In: Solomon S, et al. (eds.) Climate change 2007: the physical science basis—contribution of working group I to the fourth assessment report of the intergovernmental panel on climate change. Cambridge University Press, Cambridge